White Hot Light

White Hot Light

TWENTY-FIVE YEARS IN EMERGENCY MEDICINE

FRANK HUYLER

HARPER PERENNIAL

NEW YORK • LONDON • TORONTO • SYDNEY • NEW DELHI • AUCKLAND

HARPER ● PERENNIAL

HarperCollins books may be purchased for educational, business, or sales promotional use. For information, please email the Special Markets Department at SPsales@harpercollins.com.

FIRST EDITION

Library of Congress Cataloging-in-Publication Data

Names: Huyler, Frank, 1964- author.
Title: White hot light: twenty-five years in emergency medicine / Frank Huyler.
Description: First edition. | New York: HarperPerennial, [2020]
Identifiers: LCCN 2019054525 (print) | LCCN 2019054526 (ebook) | ISBN 9780062937339 (trade paperback) | ISBN 9780062937353 (ebook)
Subjects: MESH: Huyler, Frank, 1964- | Emergency Medicine | Personal Narrative
Classification: LCC RC86.7 (print) | LCC RC86.7 (ebook) | NLM WB 105 | DDC 616.02/5—dc23
LC record available at https://lccn.loc.gov/2019054525
LC ebook record available at https://lccn.loc.gov/2019054526

ISBN 978-0-06-293733-9 (pbk.)

20 21 22 23 24 LSC 10 9 8 7 6 5 4 3 2 1

For Colin

Contents

I

II

White Hot Light

I

I don't think we are ready to die, any of us,
not without being escorted.

J. M. COETZEE

THE BOY

W HEN THEY BROUGHT HIM IN, HE WAS ALMOST ALIVE. He was a handsome boy, in his early teens, and I saw this first because he was nearly unmarked. He was brown and slim, and we stripped him immediately, and he lay still under the lights.

The gunshot wound itself wasn't bleeding, and it stared at us from his chest. A little blue circle, an open eye.

I thought of my son. I thought of him immediately.

The young surgery attending was there early, thin and lean and hungry, well trained and arrogant and fast, and he has economy and grace and he looks through me as if he doesn't know who I am. I don't care. I'm indifferent to him. But he is sharp and hard and quick and sometimes that is what you need.

He tried to save the boy.

"When did you lose pulses?" he said to the paramedic.

"On the way in," she answered.

So he acted, right then, without waiting for anything or anyone. There was beauty in his ruthlessness.

$\bullet$ $\bullet$ $\bullet$

He stepped forward, poured iodine on the skin, put on a pair of sterile gloves, took the blade in his right hand, and opened the boy's left chest with a single astonishing stroke from his sternum to the bedsheet. Flesh parts to a scalpel effortlessly, like the wave of a hand.

"Open the thoracotomy tray," he said, over his shoulder, as the others arrived.

I was standing behind him. I could see exactly what he was doing, because the light was bright, and lit up the tissues.

The wound was bloodless. It should have come alive with little points from the capillaries, but it was clean and dry, and so I knew.

He took the scissors and inserted one of the blades between the ribs down low, by the bedsheet. Then he lifted his hand up to the sternum again, and there was the lung, pale and gray, with a few yellow beads of fat clinging to it.

By then the tray was ready. He put in the rib spreader and turned the crank, and the boy's rib cage opened like a flower.

Dark blood poured from his chest to the floor.

$\bullet$ $\bullet$ $\bullet$

When I was young, and still a student, I once went to my advisor's lab. He was doing basic research on cardiac arrest, trying to find ways to keep the brain alive for a few more minutes.

Ambition is so vulnerable to the passage of time. We remember it, we can even still feel it, but there was no solution that could be injected into the vessels that would preserve brain function, no technique that would revolutionize CPR or raise the dead with their memories intact. Instead, there were dozens of dogs.

The dogs were from shelters, and they were young, because older dogs could skew the data. He didn't know their ages exactly, but if you know about dogs, you can guess, and be close enough.

The dogs were brought out from their cages on a leash, and led to the room. They were afraid, and trembling, as if they were at the vet's.

Starting an IV in a dog is no different than starting an IV in a human being. His graduate assistant would start an IV, and inject a sedative.

It took a few minutes. They held the dog on the table, as it grew unsteady and finally lay down, and then they gave it Pentothal that put it out forever. They intubated the dog, taping the long tube to the dog's nose, and attached the cardiac monitors, and then they flipped it on its back and shaved its chest and performed a thoracotomy.

When the dog's chest was open, they would cross-clamp the aorta, stopping blood flow to the brain. Then the measurements would begin.

My advisor was not changing their fates. They were

from shelters, and would have been euthanized in any case. But I could see that it troubled him anyway, as he soothed them on the table, and spoke gently. He seemed human then, for a few moments, before they were asleep, and humbled by what he was about to do.

• • •

The surgeon worked on, scooping warm handfuls of clot out of the boy's chest and onto the sheets.

The others were there, crowding around me, straining also to see, and the entire focus of the room settled on the beam from the surgical light overhead. The boy receded, as we looked into the field.

"The heart's empty," he said, and stopped the massage, and pulled a few more fistfuls of blood out of the boy's chest until the interior was empty and pink and still. The heart lay there before us, the lung, the pleural membrane, and it all looked clean and beautiful.

Then he surprised me.

"Let's have a minute of silence," he said, and bowed his head.

It's new, that moment of silence. It's a modern ritual. The younger surgeons do it, even as the older surgeons walk away.

So everyone stopped. The nurses, the trauma team, the X-ray tech who stood waiting in the corner, her services unrequired.

For a moment everyone stood quietly under the

lights, looking at the boy's body, and no one knew the story then, no one knew anything, but suddenly it was reverent, and everyone could feel it.

A minute can pass slowly. A minute is more than enough.

"OK," he said finally, lifting the flaccid heart and pointing with his finger. "The bullet went in the ventricle here. And then it continued through the vena cava. It's an unsurvivable injury."

So the attention changed, and it became the wound again, and then he stepped aside and let the residents in.

One by one they took their turn, feeling the slack heart, and the hole, and the vena cava, because such chances are rare, to be so close, when everything is warm and the anatomy is so perfectly clear, young and new and as crisp as the morning. In that moment, the dogs came back to me.

When they were done, I did the same, reflexively, without thinking. I hadn't been a resident in more than twenty years. But I forgot all of that, forgot that I wasn't young anymore, forgot myself entirely. I just saw the wound, and felt the same cold curiosity, and was drawn in also.

I held his heart in my hand just like the others, and felt for the path the bullet had taken, just like the others. But it is an impossible intimacy, and you can't escape it entirely, and I didn't escape it at all. I saw his face beneath me—his face, that handsome boy—and suddenly it came to me with a terrible vengeance. I did think of my son again, and of a stranger, a man my age, doing this to his body for no good reason on this earth.

He was a child, I thought later, driving home. Have I learned anything?

HAIL

A FETAL MONITOR SOUNDS LIKE SURF IN THE DISTANCE.
Strapped to the woman, it crashes and hisses, because the
microphone is sensitive. Every movement is a roar, and
every drawn sheet is a waterfall.

But all of that is noise. It's the center you listen for, the
delicate *wish wish* of the child's heart. It comes and goes in
its faintness, like a hummingbird, and often the straps must
be adjusted to find it again.

The mother's heart beats also. But it's deeper, and slower,
and you can hear the difference.

The mother's heart is noise as well, her clenched hands
and her gasping and sweat—all of that is wind in the trees.
You put your palm flat. The muscle gathers beneath your
hand, and turns to stone.

That is when you listen most closely. The child's heart
can't slow down too much, or for too long. It must be
steady, and rise and fall with modesty, as if it is untroubled,
and ready for the life ahead.

I listened all day, and then night fell. Headlights came on in the streets through the window.

In the dark, the monitor sounds musical and rhythmic, like a chant, like driving a long distance alone at night when you're tired, listening to static on the radio, watching the center line flick by. It doesn't soothe you or comfort you. It keeps you awake, and keeps you looking forward. You think about your own life also, the path you are on, and the path of the woman beside you, and how you've joined elemental patterns and must follow them into the dark without the slightest understanding of how they possessed you, realizing only that they were with you all along.

Childbirth is terrifying for men. Pain at its most blinding sweeps the fears aside. But the fears are terrible also, and you endure them also, thinking about the hopes and the dreams, and the room waiting in the house, and the face on the ultrasound screen that seemed so alive and so ghostly at the same time, as the hand drifted up in the fluid beside it, and you saw the fingers curl.

That night they turned the volume just low enough to hear without straining. It receded into the background, steady and endless.

I was lying beside her on a cot, I was looking at the ceiling, I wasn't sleeping, and I was not used to being part of the world in this way.

At times she would sleep as well, at times lie awake and gasp. At times the nurse would come by; at times the doors would close and open roughly in the hall; at times someone would laugh at the desk, because there were others around us.

Beside me, her pain continued, in cycles, in gasping waves, faster, then slowing again. She didn't cry out, or turn to me, or turn away from me. I was a fixture in the room, but not a witness. Only a small part of me was a witness, and that's what I remember.

Then the anesthesiologist arrived, and the lights went on. Time for a little mercy, the wave of the wand.

"How are you doing, sweetie? Having some pain?" she asked.

"Yes," my wife replied, softly.

So I watched as my wife sat upright, with her hands on the pillow on the table. Her back was lit up in the spotlight. I could see the tips of her spine, and the spaces between them open as she bent forward, and bowed her head, and the anesthesiologist painted a perfect brown circle of iodine in the center of her back.

The anesthesiologist was steady. Her movements were deft and unhurried and exact, and I saw her experience in them, and they comforted me. She advanced the needle, and spoke in the voice of a mother.

Then I saw the flash of spinal fluid into the syringe, clear as glass, like the purest and lightest of oils. It's not quite water, and you can feel the difference in your gloved finger as it drips from the needle and gathers poised to drip again.

The final details escape me now. The procedure is the

idea of a procedure to me now, the imperfect memory of lights and needles and a white tray on the table, and the blue drape, and then the anesthesiologist was done, and the catheter was in the epidural space, and the lidocaine was flowing as another contraction began.

Lidocaine is grace. It allows thought to return, and reduces the blinding light to a dull gleam.

I thanked the anesthesiologist. She smiled and left the room, moving on down the line.

"I want a Coke," my wife said, sometime later, sometime very early in the morning.

I asked the nurse, like a supplicant, in the hallway.

"Sure," she said. "It's going to be hours."

So I walked out of the room and into the bright lights of the hall, and I felt strange, and I knew I shouldn't be so afraid, that everything would be fine, that our house awaited us, that the room was ready, that we had the money, that we had the job, that the future had come and was right to do so.

I went down the elevator, into the hospital lobby, and out the front doors into the open air, where a Coke machine sat blinking in the alcove in the wall, red as Mars, in the cold, because it was cold that night, and there had been rain earlier, and I'd left my jacket upstairs, and was shivering, and didn't care. I was alone as I fed the dollar bills into the machine and heard the rumbling descent of the Coke, and then I realized that I was thirsty also and got another.

Just then it began to hail, as it sometimes does in New Mexico at that time of year, and the hail came from nowhere, perfectly white, stinging my exposed arms, leaping off the street before me, and the paving stones, and the railings. For a moment I took shelter beside the Coke machine in the alcove, standing beside it in the narrow gap, feeling its cool red presence beside me, listening to the hail roar around me, and that, too, felt like an annunciation, as if the heavens were opening to our significance. These are the sorts of things you think at such times. Your struggles are the world's struggles; your fates are significant, the lives of your son and your wife, your husband and your daughter—they matter to creation. We see the patterns before us, and divine them, and place ourselves within them.

I knew better, of course I knew better. But I watched the hail anyway, in its beauty and transcendent intention, and I thought, *Somehow this is for us.*

When the hail stopped after just a short time, as hail so often does, I left the alcove, shivering, with an icy can of Coke in each hand. My hair was wet, and the wind through my shirt was fiery, and I felt enormously alive and breathtaken and small, and then I went back inside, and up the elevator, and down the hall, and into the room, where the sound of the monitor met me again.

We drank Coke together in the dark for a while. I put one of the thin white blankets around my shoulders. The needle had released her, and after a while she slept.

He was born in the morning.

WAR

THE FATHER CAME AND GOT ME.

"He just vomited again," he said at the doctors' station, a man my age, heavy and drawn. His son had arrived by ambulance a few minutes before, and had been wheeled to a cubicle. His chart had just been placed in the rack.

When I walked into the room, his son was lying flat on the gurney, gurgling, with a green thread of mucus stretching from his mouth to his chest. His eyes were open and blue and swept back and forth. His face was expressionless. He should have been choking and struggling. Instead he lay there, incapable and empty.

"He was hit by an IED in Iraq," the father said, emotionless. "His brain is full of shrapnel. He started vomiting yesterday. I've been up with him all night."

The muscles of his arms were thick and heavy. His body was still that of a marine. But I could see the pink half-moon of his craniotomy scar through his blond brush cut as he vomited again—another mouthful spilling down his chin and dripping to his chest.

The alarm began to ring. Within a few seconds his lips were blue.

I stepped out into the hall, called for help from the nurses, then ran back to the bedside and lifted the head off the gurney. I fumbled with the suction on the wall. He began sliding sideways off the gurney, and I grabbed a fistful of hair with my free hand to hold his head upright. The suction hissed, but his teeth were clenched tight. I couldn't get the catheter between them. I shouted for help again as the alarm rang on. I tried to suction out his nostrils, but the catheter was too large. So I stopped and sat him up as best I could, holding the oxygen mask against his face as one by one the nurses arrived. I remember the feeling of his body against me, rubbery and cool. The smell of vomit was everywhere.

As we wheeled him to the resuscitation room, I broke away for an instant and spoke to his father.

"Do you want everything done?"

For the first and only time I saw pain in his face.

"I guess," he said. "He's only been back for a couple of months."

And that was it—a brain full of shrapnel, a fistful of metal cast by an unseen hand somewhere in Iraq, somewhere on a road, or in a field, somewhere on patrol, or somewhere in a crowd, on another day when nothing should have happened, and instead this had happened.

A few months later, he was back home with his single father and a Purple Heart. There was far more to the story, of course. But I never learned it.

His father followed us down the hall. But at the threshold, I asked him to stop and wait in the consultation room.

● ● ●

For a few seconds we tried to breathe for him with a bag valve mask, but every squeeze of the bag forced more vomit into his lungs. We suctioned him again, from the nose, with a thinner catheter, and still there was more fluid rising from his stomach into his lungs. His teeth remained clenched. His eyes remained open, on their slow roll, back and forth. There was no other choice, so we pushed the drugs.

The drugs are indiscriminate. They paralyze every muscle in the body.

But as his muscles relaxed and he stopped breathing, a new wave of fluid flooded up into his mouth. I held one suction catheter, and the resident held another, as she inserted the blade of the laryngoscope between his teeth and lifted his jaw.

How had his stomach gotten so full? I remember thinking that. He was paralyzed then, and she struggled with the blade, unable to see his vocal cords, unable to see anything. I stood beside her, suctioning, pushing down on his throat with my free hand, trying to compress his esophagus.

"I see cords," she said, urgently. "Give me the tube."

I handed it to her as I kept suctioning. For a brief moment, as she slid the tube into his mouth, I thought we might be OK.

But his heart slowed as if it had hit deep sand, and then it stopped. He had been without oxygen for too long, or for just long enough.

An instant later I could feel the tube beneath my fingers entering his trachea.

"I'm in," she said as she blew up the cuff, and we attached the bag to the tube and began ventilating him. Oxygen flowed directly into his lungs. I could hear it there as I listened with my stethoscope. But the complexes blinking on the monitor were wide and slow, like afterthoughts.

So the ritual of the code began. Epinephrine, atropine, chest compressions. More epinephrine, bicarbonate. At first, for a few moments, I expected him to come back, because he was young, and his heart was strong. The alarms rang. The tube was in. We were breathing for him, compressing his chest. We followed the algorithms exactly.

Even as we continued, even as I stepped back and watched, I knew that this was the best for him. He knew nothing and no one. But he wasn't yet wasted and contracted, not yet wizened and thin and inhuman, as inevitably he would have become. His body had not caught up with his mind. He looked young and strong and nearly undamaged, with that barbed-wire tattoo on his bicep, and his strength without purpose.

On we went, for twenty minutes, then thirty. It was no use. I should have stopped much earlier.

Finally I gave the order, and everyone went quiet. He lay there, blue and still and dead, his arms hanging off the gurney. The nurse turned off the monitor.

● ● ●

I took a few moments to compose myself. I've rarely done this. What remained of his life had been in our hands—my hands—and had slipped effortlessly through them. I should have done better, I thought—I should have called the anesthesiologist, or I should have tried to empty his stomach with a nasogastric tube—I should have, I should have. But I hadn't. I'd simply walked in too late. Five minutes earlier, and he would have been alive. I knew this, and yet still I felt a terrible, excoriating sense of failure pass through me.

The father sat alone in the consultation room, waiting for the news. I braced myself, and then I went in and told him.

He looked up at me with a kind of dismal relief. He thanked me for my efforts. He declined my offer of a priest. He was composed; I was the one who was shaking. I asked if he wanted to see his son, and he shook his head, because he was done at last, and had understood the truth for a long time.

As I looked at him, I understood that it wasn't my story, and that I had no claim to it. Yet we can't help ourselves; we go on making some part of them our own.

The nurses moved the body to the decontamination

room, and housekeeping mopped the floor, and the techs put all the equipment back in place. In just a few minutes, there was no sign of the father, or the son—just the resuscitation room, waiting, shiny and clean and ready once again.

I never learned his name.

THE GOOD SON

"He won't tell us what band he was in," he said, at shift change. "The nurses think they were big, though."

We were on rounds. I was picking up his patients, so he was doing most of the talking. Cube 2—we're waiting for the ultrasound, probable gallstones. Cube 4, eighty-eight-year-old woman, pneumonia, admitted. Antibiotics? I ask. Yes, she got them. And so on. I made notes on a clipboard as we walked.

At cube 6, he paused outside the curtain.

"Here he is," he whispered. "He's sick and needs to come in."

Then, instead of continuing on as we had for the others, he parted the curtain.

"This is the doctor who is taking over for me," he said, in his white coat, to the figure on the bed. "I wanted to introduce you."

• • •

The figure on the bed was tanned a golden brown, with shoulder-length blond hair and piercing blue eyes. He looked up at me and smiled, his teeth perfect and white. A small man, I realized as I studied him, hardly larger than a child. The deep blue of his irises leapt out of a yellow background—his sclera, his face and hair. I could see his pulse in his neck. He extended a hot, stick-thin hand, and I shook it.

"Nice to meet you," I said. "If there's anything you need, please let me know."

"Thank you," he said, softly. "Some ice chips would be great."

"Tell me what you find out," my colleague said on his way out the door.

A few minutes later, I brought him his cup of ice. Small talk—how are you feeling? When did this all start? Would you want to be resuscitated if anything unexpected happened? What band were you in?

He looked away.

"I'd rather not say," he said. "If that's all right. It was a long time ago."

"In the eighties?"

A reluctant nod. A moment passed.

"I'm paying for it now," he said.

"What instrument did you play?"

I knew I should leave it alone.

"Guitar," he said. "Mostly rhythm. Sometimes a little lead."

He had a faint accent. British, or possibly Australian.

His neck pulsed. When he turned his head, a tiny diamond earring flashed in the light.

As he spoke, I could picture him easily, as a young man, floodlit on the stage as the crowd held up their lighters in the dark. So I looked at him carefully, long enough to be certain I didn't know him.

●　●　●

The residents, for the most part, were kids in the eighties, but I asked them anyway.

"See that guy? What eighties band did he play for?"

So they walked past, one by one, and glanced casually into the cubicle before circling back with their guesses.

The medicine attending thought about it, too.

"Definitely not country," he said, looking at the figure on the bed from across the ER. "But not heavy metal, either. Pop-rock, I'd say. Did you look him up on the internet?"

"Nothing under his name," I said. "But he could be using a pseudonym. He doesn't want to talk about it."

The medicine attending was in cube 6 for a long time, much longer than usual. I could see him answering questions at length, with care. I saw him shake the man's hand, and smile, and gesture to his resident.

"Did you find out?" I asked when he was back at the doctors' station.

"He won't tell me, either," he said. "Maybe the nurses upstairs can get it out of him. I'll let you know. He looks familiar, though, don't you think?"

• • •

I checked on him several times during the shift. He rose out against the others; he had significance. I looked up his labs, and entered them carefully in the computer. I called the medicine resident again and asked her when his bed would be ready. Through it all he lay still, under the bags of saline and antibiotics, without complaint, his blue-and-yellow eyes open to the ceiling.

Hours later an old woman appeared. She stood next to the bed and held his hand, and when I went into the cubicle, I was surprised to find her there. I asked who she was.

"I'm his mother," she said.

I didn't expect that. I didn't think of rock stars as having mothers. They had children, of course—lots of them. But not mothers, and certainly not an old woman like this, blinking in the fluorescent lights through thick glasses, clutching her purse. She was a tiny woman, barely five feet tall, thin also, and when she looked up at me, I saw the resemblance between them.

"Oh," I said. "Well then, do you mind if I talk to you in private?"

She patted her son's hand, and followed me down the hall.

I asked all the usual questions—how much he still drank, and whether he lived alone, and whether there was anything he wasn't telling me. And I asked her what band he was in.

She sighed.

"He played in some bands a long time ago," she said. "But he was never a big rock star. He just likes to say that. He says it makes people treat him better."

"You mean it's not true?"

"Well . . ." she said, looking down. "He exaggerates."

"So he isn't British, either?"

She shook her head. She hesitated.

"He's a good son," she said. "He's always been good to me."

"I'm sorry," I replied, regaining myself. But it was too late. I felt betrayed, and I felt revealed.

● ● ●

It must have been his life's dream, of course, but that only occurred to me later.

"He's not a rock star," I told the medicine resident, finally. "He just says that because people treat him better."

"Well," she said, looking up from the orders she was entering in the computer, "it works, doesn't it?"

He knew us well. We'd brought him ice chips, listened carefully to his questions and answered each in turn, when all along he was just like the rest, no different from the others who lay around him, or waited in the lobby, or lay in the hall on their stretchers.

Then my shift was over, and it was my turn to sign out to the next doctor in the unbroken line.

"End-stage liver disease," I said, when we reached cube 6. "Not a transplant candidate. Febrile. Admitted to medicine."

"Antibiotics?" she asked, taking notes.

"Yes," I said. "He got them."

THE WEDDING PARTY

HER FACE HAS DISSOLVED A LITTLE NOW, AND I CAN'T SEE her quite so clearly. It's only her face that has begun to resist me. Why, I do not know. Maybe it's the desire to turn away.

I remember her nature better than her face.

The school was in Japan—a boarding school where my parents were teachers and I was a day student. She lived in the dormitory. Her father was an American journalist working in Beijing. There weren't any prep schools in Beijing then. So he sent her there.

She'd drink a Coke every morning before school. I'd arrive from the train station with my parents and brother in one of the immaculate taxis whose drivers wore white gloves, and there she'd be, in front of the dorm.

The school was set high on a hillside overlooking the city. My family and I would walk past the dirt soccer field, past the dorm, then up the stairs to the campus and our classes. She'd roll her eyes at me as we passed, sipping her

Coke in the Japanese morning. It's not an easy thing to go to high school with one's entire nuclear family on display, just as it is not an easy thing to go to high school without any family at all.

Japan has English weather and sea air. The cities are gray buildings and wires and telephone poles, with hardly any greenery in sight. I was cold a lot of the time, cold at school and cold at home, in our house without insulation by the train station, with its tiny yard and tatami floors. Taking a bath or a shower in winter was misery: the bathroom icy, the water pressure low. And so I was often unwashed back then, my hair limp, my clothes a little ragged. But there were good showers in the dorms, so she was always clean and pink, dark-haired and brown-eyed and superior, as she surveyed the arrivals from her perch on the steps.

We took honors English together. Our teacher was, I believed at the time, batty and old. She made us read *Hamlet* aloud and act out the parts. She made us read the chapter introductions to *Tom Jones* instead of skipping over them like the other English class did. She asked us why we thought Queequeg's coffin saved the narrator in *Moby-Dick*. She asked us what *Bartleby, the Scrivener* might say about the Japanese work ethic and the endless stream of identical suits that filled the trains every morning. She asked us why *Waiting for Godot* might be anti-scientific. She took us on field trips to Kyoto, which had nothing to do with English literature, and made us eat traditional

Japanese meals sitting cross-legged on the floor in tiny restaurants in the old city.

Getting into college consumed us. This school, that school, all those shiny brochures, with leafy quadrangles and ivy-covered buildings and beaming, attractive faces, all those portraits of keen-eyed professors and the provocative classes they taught.

Now my alumni review follows me from address to address wherever I go. The faces are exactly alike, the smiles just as blinding in their enthusiasm, the professors every bit as provocative as they were in my time. The grass is just as green, the oaks just as broad, the fall color on the hills no less scarlet. There is that same church steeple also, and that same endlessly photographed admonition on the plaque:

CLIMB HIGH, CLIMB FAR
YOUR AIM THE SKY, YOUR GOAL THE STAR

Back then, for the two of us, poring over the pages, those places seemed like the promised land. How exciting, all those claims of excellence, and we swallowed the vision whole.

She chose an Ivy League university, and I chose a liberal arts college, and once we even kissed for a while after everyone drank too much at a party, a few days before we each left Japan for our new lives on the East Coast.

We talked on the phone sometimes during our freshman year, lonely, far from home, and I visited her once, in the spring. But I went to see her as a friend, and I left as a friend, because truthfully I never thought of her as anything else.

A few months before, over winter break, she'd gone to South Africa, and had liked the accent enough to acquire it. For my part I had worn, foolishly and proudly, a jean jacket covered with antinuclear buttons to go along with my political convictions.

So there we were with our affectations—my ridiculous jacket, her even more ridiculous accent—walking around the campus of Brown University, full of the usual aches and hopes and arrogances of college kids taking their first steps into what they have been assured will be the brightness of life.

Providence was rough in the eighties. We were walking in the darkness back to her dorm when a car pulled up beside us and the townies inside started shouting obscenities and throwing empty beer bottles. We were afraid at first, but they were drunk, and their aim was off, and I can still hear the glass shattering on the brick wall of the building beside us as we ran down the dark street between the pools of streetlights. They roared away, and she said something in her fake South African accent, and we began to laugh.

I remember lying on her roommate's bed late at night, listening to Boy George sing "Karma Chameleon"—a song that makes no sense—and every time I hear it on the radio I think of that weekend, of how young we were, how excited and uncertain, how confident and afraid, how vulnerable to the years ahead. I remember going to a diner the next day, on a gray and rainy day, eating fried blueberry muffins soaked in butter and watching the cars pass through the wet streets. I must have taken the bus to see her. I can't remember how many days I stayed, or what else we did. But I remember the diner with perfect clarity, smoky in the morning, the blueberry muffins shining with butter on the table, and all that rain.

●　●　●

She sent me an email. We had lost touch by then; I hadn't spoken to her in many years. She was organizing our twenty-fifth high school reunion—LA, on the beach, a good midway point since so many of our classmates were in North America and so many were in Japan. She had gotten my address from the school. Phone me if you can, she wrote, with her number, I live in Canada now, and I thought of her South African accent again.

So I dialed the number. I remember being nervous as the phone rang, wondering what I would say, how I would sum up the years for her. I felt on the spot somehow, wondering if we would be awkward with each other.

When you get to a certain age, reunions start to mean something. Five- or ten-year reunions are only warm-ups for the real thing. But a twenty-five-year reunion is a different matter. The world has had plenty of time with you, and if you're coldhearted, you realize that unless you're very lucky, or very unlucky, the trajectory on which you find yourself is the trajectory on which you will remain. The path is worn too well, the choices have mostly been made, and the end, for the first time, begins to take shape in the distance. You meet old acquaintances again, you tell your story, and they tell theirs.

Telling that story is not an easy task if you take it seriously. The temptation is to not take it seriously, and to roll your eyes at those who do. You can laugh it off— *I did a little of this, a little of that, isn't this all just a game anyway?*—or you can claim fulfillment—*Life is what you make it. I've made it fine.*

But ultimately it's impossible to avoid assessing your past. As you get older, the past rises, in fragments, spilling into your thoughts, catching you out at odd moments. The fragments do not fit together. They resist synthesis, and what leaps out is not clarity but power. Not so long ago, the world awaited you, and your task was to enter the abundance of your future with all the effort you could bring. Even now you cannot accept an irregular line, one that simply starts and stops, as if without significance.

●   ●   ●

"Huyler," she said, as she always had, wry as ever, having looked at the caller ID.

She was a professor, an archaeologist. Her specialty was the Roman settlements of Cyprus. She said she loved her work, speaking for once without irony. She loved going to Cyprus in the summer, to the ruins and excavations, and the perfect gold coins she unearthed, unspent since the time of Christ. She was a numismatist, a scholar of coins, and the coins were found mostly in graves.

The girl on the steps of the dorm with her Coke—this was her path. How unlikely it seemed; how vulnerable to the disdain of the practical world.

Yet part of me envied the purity of her work. The graves were fascinating, after all: knowledge for its own sake, knowledge with only the faintest application to our lives on earth. The graves and coins had so many stories to tell, and there were so many worlds beneath our own, there in the ground at our feet.

I didn't go to the reunion. I had no excuses. I regret turning away; I regret not wanting to look at all the time that was gone. She emailed me pictures. And there she was, with her straight brown hair, her slightly crooked teeth, wearing sunglasses, on a pier in LA with the others—the same girl, looking only a little bit older, with her partner at her side, though he wasn't in our class. He was a few years younger and had lived in the dorm as well. I must have passed him hundreds of times, yet I didn't remember him at all. I stared at his face and felt not the slightest flicker of recognition.

* * *

Dear Friends, she emailed only a few months later, to everyone in her address book, under the heading "Bad News."

I've been diagnosed with acute lymphocytic leukemia. I am beginning treatment now, and am optimistic about my recovery. I am blessed by the support of many friends, my family, and my partner.

As soon as I saw her message on the screen, I knew she was gone. And so I sat there for a few minutes before picking up the phone and calling her.

Acute lymphocytic leukemia is curable in childhood but lethal later in life. She had won a dark lottery. As the details emerged for her, the news grew worse. The cytology was unfavorable. The best one can realistically hope for with a diagnosis like that is a remission or two, a few years of grace. But the oncologists had told her she had even money, as they do.

So we began to talk regularly again. I was the only doctor she knew as a friend; a buffer, perhaps, between worlds.

"We'll just have to see what happens," she said that day on the phone. *We'll just have to see*—as if she were not alone, as if she could step back and watch. She'd just bought her first new car, a Honda Civic, which she said should last for a long time, according to *Consumer Reports*. Her professional life was books, students, gold coins, and summers in Cyprus, with its beautiful light and dinners with friends, the Mediterranean everywhere. My professional life was car wrecks and shootings and drunks and overdoses and the other darknesses of the ER. I knew what she was facing in a way that she did not, and I could hear that difference in our voices.

Months later, when she understood better, she told me about the skeletons she had uncovered, and how they

showed the labors of the body—certain injuries to the
fingers so characteristic of fishermen and their heavy nets,
or patterns of wear in the sockets of the hips. One could
learn a great deal about a person sometimes, she said, and
someone looking at her skeleton would one day be able
to do the same.

I've never seen a skeleton in a grave, and so her point
would not have occurred to me. But when she spoke of her
skeleton, and what it might reveal about her, I don't think
she wanted me to reply. I think she wanted to give voice to
her thoughts, to consider herself and her fate in the context
of history, as another human story in the chain, as if she
were part of something larger, and greater, than herself.

I didn't know what to say to her on that first day. It is
one thing to talk to a patient, and it is another to talk to
a friend from the past, your own age, an equal in every
way. My heart was in my throat, and there were tears in
my eyes, which I am unused to.

I told her she would go into remission.

She laughed.

"We'll just have to see," she said.

• • •

Chemotherapy kills millions of cancerous cells. But every
last one must die, and usually a few escape. The cells that
survive are more likely to be resistant to further chemotherapy. They keep on dividing and return in numbers,
stronger than before.

So the best chance for survival is a bone marrow transplant. Very high doses of chemotherapy are given, killing all rapidly dividing cells and wiping out the bone marrow.

Then new marrow from a donor is injected into the bloodstream and finds its way like magic to the bones.

If you are very lucky, the protocols work. But they are scorched earth, and bring you close to death. You lose your hair. You vomit. Your mouth fills with bleeding sores. For a time you have no immune system at all, and an otherwise harmless bacterium or virus can kill you. So you cannot get too close to other living things. You can't be around flowers, or grass, or children. Everyone who touches you must wear gloves and masks and gowns. You must hope the transplant takes, and that there is no leukemic cell left behind, like a single star, waiting to come back. So you sleep, in isolation, exhausted to a depth that few of us can ever understand.

From my days as a medical student I remember the silence of those units, where the doors were always closed, and I remember the peculiar sense of the abyss, more than anywhere else in the hospital. Everyone was waiting—waiting for the marrow to take, waiting for the counts to rise.

And so I didn't hear from her for many weeks. Her bones were empty, and her brother's blood was flowing in her veins.

● ● ●

The transplant worked for a while. They usually work for a while. In the respite, when she was back in the everyday world, her boyfriend asked her to marry him.

I didn't know him. He was a blank space. But his love for her was plain to see, as he drove her to and from the hospital, as he brought her popsicles and smoothies when her mouth was full of sores, as he posted encouraging updates on her cancer blog. They are one of the new rituals of the modern world. I never sent an email to her cancer blog, but I did light her a candle in a French cathedral, on vacation.

Once, I asked her if she felt that she'd learned anything from her ordeal. We want to believe that we learn things from ordeals, after all.

"No," she said, and laughed. "Not a thing."

It wasn't what I wanted to hear—suffering like that, surely, must teach something.

She was in remission, and she thought she had a chance. That is just one of the many terrors of cancer: its silence, allowing the mind to try to spin its way out. It is hard to believe in black numbers on a page. It is even harder to believe them when all the ordinary details of life continue as they do—papers to grade, students to see in the offices. And a wedding to plan for.

When you choose to get married, perhaps it is that much harder to believe you are going to die. Perhaps that is one of the reasons to get married, after all. Or maybe it is a statement of another kind, a form of defiance. Or maybe it is a form of resignation. Or maybe it is something else entirely.

She sent the invitations to everyone she knew.

●   ●   ●

So I flew to Canada from New Mexico for the wedding. I rented a car in the city and drove through the darkness to the hotel for the rehearsal dinner. I got lost and arrived late.

The hotel was large, faceless and generic and disappointing. I wandered around the lobby for a while, seeing no one I knew, before I finally went to the desk to ask. *I'm looking for a wedding party? A rehearsal dinner? I hope this is the right place.*

I found them in the bar—a few dozen unfamiliar faces, standing, talking, drinks in their hands.

And there she was. I had not seen her in more than twenty years. I had not seen her since we were young, and I had a little son by then, tucked neatly in his crib thousands of miles away.

She was walking toward me, smiling, her face unlined and unchanged. Her hair had grown back, brown as before. But she had gained at least fifty pounds, swollen from the prednisone. Another body had taken the place of the one I remembered.

"Huyler," she said as we hugged briefly, "thank you for coming."

In that moment I could feel how much time had

passed, and I suddenly understood the limitations of my role, in a room full of strangers, the friends of her adult life. I felt like an intruder, and I wondered if I had imposed in coming all that way.

I told her it was nothing, and that I had very much wanted to come. It was true, but even more true was that I didn't want to be the person who didn't come.

"You look good," I said.

She rolled her eyes.

Time, so far, has treated me well. My hair isn't gray; my weight is unchanged. My age only becomes clear up close—the crow's-feet around my eyes, the skin of my hands.

"You look practically the same," she said. "It's disgusting."

I stayed for a while, had a couple of drinks. I was tired from work, from the long flight, from the struggle to find the hotel in the dark. She was having dinner with her family and close friends, and I wasn't invited, and I realized that I didn't want to be invited. I wanted to go back to the cheap hotel I'd booked online and sleep.

● ● ●

The wedding took place at a beautiful hotel. A river flowed through the grounds; there were pines and lush northern grass and granite boulders, gray with threads of blue. From the restaurant you could stare at the sheets of slate-gray water flowing over the rocks and coming to a boil at the chute for the waterwheel.

The crowd was much larger than the one the night before. I was not alone in the distance I had traveled; there were guests from England, from Cyprus and California and Japan, dressed in their Sunday best. Several high school classmates had brought their wives and children. And there was our high school English teacher.

She looked terrific, healthy and spry, the kind of retiree I could imagine hiking up a mountainside with an aluminum walking stick. She didn't look old to me at all as she approached, beaming, to give me a hug. She was a healthy seventy at most. And so she had been only in her forties when I was her student, not at all old in the way I had assumed.

An English minister presided over the wedding. The groom's family had once been members of his congregation and had asked him to conduct the ceremony. They had explained the circumstances. He had a high position in the Church of England, with many administrative responsibilities, but had flown across the ocean nonetheless, and accepted no payment.

The string quartet began to play.

She was being driven to the hotel in an ancient Rolls-Royce, and she was late. The minutes passed, and people began to shift in their seats. The groom stood waiting in his suit. He looked at his watch. A woman came up to him and whispered in his ear. Then he and the minister exchanged a few words in the doorway. The minister nodded, then approached the podium.

"I'm afraid there's been a bit of a problem with the

car," he said, with a smile, as the crowd grew quiet. "It's a
Rolls-Royce, and so of course it's broken down. But they
will be here in a few minutes, and I've been told we should
begin."

And so he began.

I have no faith myself, but like so many unbelievers, I
envy the faithful. He did not read from prepared remarks.
Instead he looked out at the audience and spoke, as if
without effort, and from the first words he did not skirt
the truth, as a lesser minister might have done. Instead he
addressed her illness directly. He spoke of pain that resists
understanding and the grace that can emerge from terror
and darkness if we allow it to. He spoke of the redeeming
power of love for one another, which flows from God,
and which alone permits us to endure the mystery of suffering. But, he said, we must choose. Faith is not a gift.
Faith is up to us.

I had heard the message before, but it was nonetheless
a balm, and he cast a spell as only the best speakers can.
I could feel his words settling over the crowd as he con-
fronted what remained unspoken in the small talk and the
silences, as he acknowledged the trials of faith head-on, as
those of faith, if they are honest in their conviction, must
do. And even for me, for a little while, he crossed some
of the distance.

When he drew to a close, she was there.

● ● ●

Later, at dinner, I sat beside my English teacher and two archaeology professors from a famous university—elegant people in their sixties, a husband and wife. The woman had clearly once been very beautiful; she was striking still, with her shock of white hair and her French accent. What a small world theirs was, I thought, listening to the gentle academic talk, who was where and so on, over the clink of silver—an easy murmur, like a radio turned low.

My English teacher, as was her way, began asking cheerful questions of everyone. Her hair was newly cut, and she was wearing a vaguely Japanese outfit. I saw her intelligence as if for the first time. I saw her batty enthusiasm again. But most of all I saw her kindness, and it suddenly felt very nice to sit there beside her, after so many years, watching the river flow by just outside the windows while the waiters brought us wine and filet, or sole if we chose.

The bride and groom sat together, alone at a table facing the room. For a moment she looked heavy and stern, in her white dress, with her hair pulled back, her face in shadow in the dim lighting. I'd had a couple of glasses of wine by then; normally I wouldn't have said it.

"She looks like a Roman emperor in a toga," I said.

My English teacher laughed. "You're right," she replied. "She does."

• • •

I flew home from the wedding in Canada the next day, and I remember looking down at the passage of the fields thirty thousand feet below—one after the other, long grids, somewhere in the Midwest. I knew I'd never see her again, and she knew the same, but by then she was used to unspoken goodbyes, I suspect, and I was simply one of many hundreds who had passed through her life in one way or another. I understood this, and neither of us made too much of it.

I felt numb, detached and tired, as the fields succeeded one another; if there was sense to be made, I couldn't see it, and if there was redemption to be had, it escaped me. The wedding, the dinner, the groom's love, the eloquence of the minister's words, the kindness of my English teacher, the tears in the eyes of the crowd, all the money that had been spent, and the grave that was waiting no matter what was said or done, no matter what toasts were made or sentiments unleashed, as everyone thought of themselves, too, and the fragilities, and their own husbands and wives and children, and the overwhelming sense of the end of hopes and the end of dreams—none of this was easy to think about, and yet I couldn't help but think about it. I'd flown all that way, after all.

When she got up to dance with her new husband, and the music played, everyone in the crowd fell silent, and many were in tears. That reverence, I suppose, is the best we can do. How thin it seemed, then and now.

The cells returned a few months later, stronger than ever.

● ● ●

Several of the nurses, she told me, were crying also when they saw her back on the ward. Her blood was filling up again. The nurses knew what this meant, and so they cried when they saw her there. And she, of course, knew what their tears meant as well.

I had a final conversation with her not long afterward. She was on her speakerphone, too weak to hold the receiver, lying on her bed at home, and she was bleeding— oozing into the bedclothes. Every few hours her husband would change the sheets. She told me this lucidly, matter-of-factly. They were running low on sheets; he had to buy more. Her words were slurred, crackling with static. Yet her mind was clear. She told me she could feel her body shutting down. She was watching herself go.

I knew exactly what was happening. They could have given her packed cells; they could have given her platelets. But they didn't because there was no point. Instead they'd sent her home with a bottle of oral morphine. I imagined her husband changing the sheets, and I imagined them dripping blood onto the floor, because I've seen that also. I imagined the washing machine running in the basement below her.

I remember stammering, struggling to find the right words. Bravery for something, I can understand. But bravery for nothing, bravery for its own sake—that resists me. I don't understand how she could lie there as she did, so lucid and calm, watching the bedclothes fill with her own blood, waiting for her family to arrive, so composed on the phone as she said goodbye, one by one, to all of her friends.

The next morning I sent her an email. I composed my thoughts as carefully as I could, and told her how much I admired her decency and courage and grace, though of course it didn't matter at all what I thought, or think now; it mattered for nothing, and I didn't expect to hear anything back, weak as I knew she was. But a few hours later I did hear back.

This is what she wrote:

> Thanks for being so supportive through this, lending an ear, and being a good friend over the many years we have known each other.
>
> Love,
> Danielle

Two days later she was gone.

●　●　●

Her husband sent the email almost immediately. She had passed away peacefully, surrounded by her loving family and friends. He told the story that decorum requires: she did not suffer in the end. He thanked her family and his own, he thanked her friends, he thanked her doctors and nurses for the wonderful care they had given her. He did his best to tie the bow.

Yet I could see the depth of his grief, and I could understand the relief he must have felt nonetheless, when the ordeal was finally over, when everything was done.

He had helped her as much as it is possible to help another person, and then, in an empty house, he was left composing the sort of lines that eventually will be composed for all of us—*he passed peacefully, she was released into the arms of the Lord, he is preceded in death by his beloved wife, she is survived by her sons and daughters. . . .*

I thought of him, the boy I never knew, my unknown schoolmate, standing in his suit, with his father as his best man beside him, waiting for her as the minister spoke, waiting as the Rolls-Royce sat smoking on the roadside, and I thought of how she had rushed in, pink and flustered with the bridesmaids, and how everyone was relieved, and how everyone was smiling. I remember how she stood breathing and alive in the doorway, looking at the crowd, just before the service began.

JEHOVAH

HE WAS MIDDLE-AGED, A DEAN AND A PROFESSOR, AND THE melanoma had spread to the lymph nodes in his groin and matted there, like seaweed. As the lymphatic fluid collected, he had a hard time wearing his pants with a belt. He had a committee meeting that afternoon and wanted to sit down comfortably.

He asked me to drain the fluid with a needle. I hesitated, but I did it because he said others had done it before.

So I put the needle into the mass in his groin, and yellow fluid leapt out under pressure, and I filled syringe after syringe until his groin was flat again.

He thanked me and said I had the touch, then put his pants back on. He was wearing a shirt and a tie, and looked like a dean and a professor, and as I cleaned up and threw away the needle and the gauze and the full syringes, I realized that no one would guess his secret for a while longer.

When we talked, he said he was going to work as long as he could, and that they'd have to carry him out in a box.

• • •

She was not yet forty and had gone back to school. She lived alone, and looked fine, but had brain metastases from breast cancer nonetheless. They brought her in for a seizure. When I saw her, she had woken up again, and knew where she was, and why. She had never had a seizure before.

She kept asking me, *Does this mean it's the end? How long? Should I call them? Should I tell them to come? Should I call them? What should I do?*

She knew it had spread to her brain but said she'd been doing fine.

I said I thought she'd be OK for now, as long as she had someone to stay with her, and she said she'd ask her boyfriend, but really she was thinking of calling them, calling her family, and I said that she probably should because she was going to need help soon, and she nodded, so matter-of-fact, without any sign of emotion at all.

• • •

She was quiet, from Mexico, and wouldn't tell her family what was wrong, but they knew something was wrong anyway, even though they lived in the US, and so her sister went down to get her and brought her back on the bus.

I got the family out of the room and stood there with the nurse, because she didn't want anyone to see, not even her sister, and when we peeled off her shirt, we understood why, because the cancer in her breast had eaten half her chest, and it was black, and reeking, and I could see her exposed ribs through the red raw tumor, which she had hidden from everyone.

"She is shy," said her sister, who spoke good English and was much younger. "She has always been so shy."

• • •

She said she was from Atlanta originally, but had been out West for a while and it felt like home now. The metastases from her ovaries felt like hundreds of pebbles, all over her shoulder and back, just beneath her dark skin, and I asked her if they were painful and she said no, they didn't hurt at all, and she was grateful for that at least.

She sat there in a fur coat and wide sunglasses, wearing a hat with a brim, and large gold earrings, laughing with her sister about something. They were both in their sixties, but they laughed like young women, explosively and joyfully.

• • •

He was in his early twenties, and I'd never seen anything like it, neurofibromas in the lymph nodes of his neck, mats of tumor, like a slowly growing vine, until his neck was drum tight and wide as his thigh and slowly choking him. I didn't talk to him, but I saw him, sitting upright and extending his neck to breathe, sent from the hospice even though there was nothing left to do. He stared up at the ceiling in silence.

• • •

A bleeding, blackened breast, a little pumper of bright red blood from an arteriole, not enough to be dangerous but enough to keep oozing through blouse after blouse, and I remember her, too, refusing any treatment save cannabis oil. She lived alone, in a trailer in the mountains near the border, refusing and refusing while the tumor grew larger. Finally it had started to bleed and made her give in, and so they sent her by ambulance from a clinic in a little town called Silver City.

I was trying to find the bleeder in the tissue, trying to put a stitch around it, but it wasn't human flesh I was working with. It was cancer, like bonemeal or wet sand, and the sutures pulled through at every knot. She felt nothing because the cancer had destroyed the nerves, but it stank up the room, and kept bleeding. I kept trying to soak up the blood with gauze, looking for the vessel as I put in stitch after useless stitch, but I couldn't see anything down there in the muck.

As I worked, she talked about cannabis oil.

• • •

He was a pilot in his forties with an astrocytoma in his spine, and the pain down his leg was finally too much. He lay sweating and writhing, and told me that he needed to give his family a rest. But then the drugs worked, and he could bear it again, and he changed his mind and wanted to go home.

So we talked a bit about flying, and how he'd flown a new jet prototype across the country when the instrument panel went dead, and had just kept going, because the sky was clear, and he knew his altitude, his airspeed, and his heading, and that's all you really need.

I remember his little wave of thanks as he limped out.

• • •

He had lung cancer and pneumonia and a neutrophil count of zero, and his wife held his hand and told him they'd get through this like they'd gotten through everything else. Then his breathing got labored and he started gasping and I asked again and his wife insisted. They were about to enroll in a clinical trial.

"There's always hope," she said. "Always. You can't take away people's hope. Please don't do that."

So we intubated him, even though the oncologists hadn't been honest, and had offered the clinical trial, as they do. It was either that or let him die right then instead of a few days later, and I chose a few days later because I had to.

As we were wheeling him to the ICU, she thanked us for saving him, and I remember that.

• • •

He was a medical student, about my age at the time, and I was working alone, and testicular cancer is curable, but not for him, he'd gone through everything, and there he was, crouched facedown and gasping on the gurney, because that was the only position that would allow him to breathe.

So I let him crouch there, and tried to give him breathing treatments while he was on his knees, because his lungs were completely full of tumor. He said he wanted everything done—*do everything, don't give up on me, please.* He said they were holding a place for him in school.

*   *   *

The young woman reading the book in the cube, sent over from a clinic when her routine white count came back, and she looked fine also—she didn't look like someone with blasts in her peripheral smear and a white count of 120,000, not at all—and I put my head in the room and asked her how she was doing and said we were still waiting for a bed to open up.

She smiled a quivery smile and put down her book for a moment, not quite teary but not far away.

"Big day," she said.

*   *   *

I can't remember when I saw her, and I can't remember how many years it's been. I know it was the new building. I could smell the odor all the way down the hall.

She was twenty-two or twenty-three, and she'd been beautiful, and you could see it even then. Her left foot was a black and reeking cantaloupe-size ball of flesh, fiery, like an image of mercury. On the surface of the ball that had been her foot, the remnants of toes.

She'd refused to let the surgeons take it off because

she was a Jehovah's Witness. She kept coming to the hospital anyway, but osteosarcoma is fast and spreads everywhere, and it was all impossible, and later it wasn't even certain what she understood about Jehovah, if she understood anything at all.

She lay there on the bed, skeletal and gaunt, and her chest X-ray was one cannonball of a metastasis after another.

She was seven months pregnant. Her belly stood out against the rest of her because the baby was alive.

I had no idea what to say to her that night. I had no idea at all. So I asked her about the baby.

"I'm having a girl," she said, and smiled.

THE MOTORCYCLE

HE KEPT NODDING OFF. WE PULLED THE CURTAIN AND watched him closely because they die so gently. They lie so still and turn so blue as their hearts keep beating. Their hearts beat all the oxygen out of their blood, and blood without oxygen is blue.

Most deaths are pale. When your heart stops, you don't turn blue.

You turn blue in hangings for the same reason. The heart beats on, without oxygen, for a while.

Drowning will turn you blue. So will some poisons, like cyanide.

But those are painful ways, terrifying ways. Heroin is not terrifying. Heroin is a soft warm night where no one cries out and no one is afraid.

"Do you want Narcan?" the nurse asked.

I looked at him. He was still breathing. The numbers were OK on the screen.

"No, let's just watch him," I said. But it was close. He was on the edge.

·   ·   ·

Narcan wakes you up in seconds. It's brutal, like childbirth. Bang—you're there, shivering and gasping under the lights. We save more people with it than any other drug.

When you see someone dying of an opioid overdose, and you give them a few drops of a clear liquid, and they stop dying, it makes you think. It makes you think about the nature of consciousness. It makes you think about free will and death.

When they come back, they don't know. They don't feel the risk. There was no cold breath. The fear is abstract.

"What happened?" they ask. "Where am I?"

So you explain. You tell them where they were found. You tell them it was only a few minutes between this conversation and the grave. You don't use those words.

"I think he needs it," the nurse said, a while later. I'd been watching him also, where he lay, head propped up on the pillows, a little silvery string of oxygen beneath his nose.

"OK," I replied.

"Point four?"

I shrugged. "Sure."

So she went to get it, and I waited for her, and we walked to the cube together.

We stood there for a moment, and she was right. He wasn't breathing well enough.

He was young and worn out. I've seen him a thousand times.

● ● ●

Heroin has a look to it. After a while, you begin to see it in the street. You see it in the clerk at the convenience store; you see it in the faces at the barbershop. You see it at the car wash, and you see it at the state fair. Life continues with it. People go to work, and come back from work, with their secret, and its many blessed moments, because heroin has beauty in it.

The beauty is what makes it impossible. The beauty is why they turn to their legs when their veins are gone and force the needle in anywhere, why they steal and cry and reduce themselves so completely.

The nurse wiped an alcohol pad over the port in his IV. The paramedics started it when they found him. It meant he still had veins, and it's a relief to have an IV. An IV is a form of communication. An IV connects this world to the next.

She injected the drug. We waited. The seconds passed— ten, twenty.

Then it hit him. He sat up, as if risen from the dead. He ripped the mask from his face.

"What the fuck," he gasped, shivering.

So I had to watch him longer. The heroin was still there, circling. He needed to prove he wouldn't die when the Narcan wore off.

• • •

It was a sunny day, windless and cool. I was riding my motorcycle down a dirt road in the desert. The mountains were to my left. The sage and empty desert was to my right. Miles away, I could see the silver line of the interstate bisecting the land.

I should not have been riding a motorcycle. I was too old to ride a motorcycle. But a motorcycle gives middle-aged men the power of youth in their wrists. You twist it, and the road before you snaps into place, and the wind pours, and you feel light and athletic again, because all motorcycles are fast, and clean, and ethereal, and when you come up to a corner and flick the handlebars out, and the bike leans in to a perfect, glorious curve, you are as young as you could ever want to be.

Foolish, foolish, that seduction. You can feel the risk around you, and you don't care.

I stumbled across some other men. They were riding in a pack through the desert. My bike is street legal, but it is at home in the desert, on a rough dirt road, at forty miles per hour. You lean in, you stand on the pegs, you streak across the sand, and there is a rooster tail in your wrist, and it's a well-made machine, and it howls if you want it to, and climbs up a hillside like a monster if the grip is good. It ticks over, also, on a cool morning when you can see your breath but are warm in your clothes, and the helmet with its visor makes it all seem far away. You are anonymous then. You could be anyone.

It's a common story. Too much speed, and other men ahead, and around the curve, too wide, into the soft ground, as the rear tire slides away and then catches on hard ground again.

I put my foot down. They teach you not to, but it's a reflex. I put my foot down, and it snapped my ankle, because a motorcycle is not a bicycle. A motorcycle weighs hundreds of pounds.

I hardly felt anything. But when the rear tire caught the hard ground again, it threw me up and over, and I came down hard on my head and shoulder. My helmet did its job, and I felt it absorb the blow and collapse a little, but my ankle was broken, and the pack continued on and left me because they had not seen me fall.

So I stood there, hobbling, in the desert, by myself, watching them disappear into the distance like little beads.

I lifted the bike. It was difficult. I had to stand on my broken ankle. The bike started, and I was thankful for it. Then I rode twenty-five miles back home, alone, across the desert.

• • •

A woman was there. She was in her forties and dressed professionally. I knew immediately that she was his mother. She was standing outside the room. She looked drawn, and weary.

I got up, and walked over, and introduced myself. She attempted a smile and tried to look at me.

"I'm trying to get him into rehab again," she said. "But he won't go."

By then we were standing outside the door. He could not hear us. She looked at her phone, then at me.

"I have to get back to work," she said. "I don't know what to do."

She looked at me beseechingly.

"How long has this been a problem for him?"

"He started with pills," she said. "In high school. He got them from his friends."

We were silent for a moment. I knew this story. I'd heard this story again and again.

"I can't force him," I said. "He's an adult."

She sighed.

"That's what they always say," she said. "But what am I supposed to do? Can't you keep him?"

"I can talk to him. And I can watch him here for a couple more hours."

"Isn't there anything else? Isn't there anything?"

I wanted to say yes. But it wasn't the truth. When you are an adult and have no money, there is nothing for you.

"I can't afford rehab again," she said. "It's just too much. I have other kids."

This also is the truth. Rehab is among the cruelest of industries. It charges what the market will bear. Insurance usually doesn't cover it, and most have no insurance anyway. When people are desperate, the market will bear a great deal. When people love their children, or their brother or sister, sometimes they take out loans.

•   •   •

I was on the table, blinking under the lights. I was warm. I was not afraid, because they'd given me a brushstroke before they rolled me in. I felt the momentum upon me. I could feel myself being carried, like a child, and I realized that I'd reached the point beyond any choice at all.

I remember the orthopedic surgeons. I remember the white vial of propofol in the anesthesiologist's hand, and how closely I watched him press the plunger. For a moment, it felt like my will against the white liquid flowing into my arm. For about ten seconds, as it burned in my wrist, I thought it was a contest.

Anesthesia is the experience of death, which is to say it is no experience at all. Absence is not a reflection on absence. Absence is not looking down upon the living, or drifting in the reaches. To be brushed aside like that, to be waved away while you are thinking, while you are concentrating as hard as you can, while you are watching the white liquid flow into your arm, is to realize that you are not discrete in the world. You are part of it. You are a delicate and transitory thing. Your mind is a feature of the world like a flower is a feature of the world. It opens and closes its petals, and receives no answers.

•　•　•

I sat down beside the bed and spoke to him: *You're going to die if you keep this up. Do you understand? Your mother is here. She had to leave work. You are not a child. You're a man. Look what you're putting her and the rest of your family through. They can't help you because you won't let them. You are hurting them, and hurting your mother. You have to help yourself. You're a young guy. You have your whole life ahead of you. What are you doing?*

Usually, there is remorse. Usually they agree. I don't know if it helps or not, that grandstanding. I know that partly I am speaking to myself. But not speaking is a colder act.

When they're neither euphoric nor withdrawing, the world is clear again. It's back in balance. But then the withdrawal begins. It distorts thought, and the struggle between the present and the future starts once more.

He looked at me. He was wide-awake.

"Can I go now?" he asked.

•　•　•

I was lying in my room at home. My leg was in a cast. I could see the blue sky and the sun through the window.

For a few days I could feel my heart beating within my ankle. But it was bearable pain.

Bearable pain is a contest that you can win if you choose to. And what you choose reveals you.

I had thirty pills. The first five I took for pain. The last twenty-five I took for their whisper of grace. I could have gotten more so easily.

It was pleasant to lie there, with a book, and my leg on the pillows, and the sun shining through the window. It was so pleasant to drift softly through the pages. A weight had lifted. The light was warm on my leg. I could see the path ahead.

MERCY

I WAS WORKING WITH BEN. THERE WERE NO RESIDENTS OR students. It was just the two of us, in our pod, on a Wednesday.

Ben had just finished his training. He's smart and steady and believes in himself. He does not reveal weakness.

Danger, for young men like him, is the North Star. There is glory to be found within it, where action can be taken and the world put right again.

Once, he convinced me to go skiing with him. We skinned up the mountain, which was closed for the season, on touring skis. There was no one there.

Up high, New Mexico changes. The desert shines in the distance, but you are in juniper pine and piñon and cedar. The sky overhead is a perfect and radiant blue.

An empty ski area has an eeriness to it. The chairs hang silently from the lifts. The sun was blinding on the snow, on the windless day.

We set off in the mid-morning, skinning directly up

the slope toward the summit. We were the only ones on the mountain.

Ben was once a mountain guide. He has blue eyes and blond hair and is close to the Germany of his childhood. I was twenty years older and half as strong. I thought he might turn it into a competition anyway, because so many young men do this.

But there was none of that from him. I noticed it immediately.

* * *

It was our turn. When the pagers go off, a little line of text follows.

I read the text and felt my quickening pulse, because it was happening again, and I didn't want it to happen. I was tired, a little low. I wanted to be somewhere else.

A few months before, Ben had been my resident. Now we were equals as we walked together down the long hall to the trauma room to get ready.

"I'll do it," he said.

I was happy to let him. I'm tired of the rise and fall, tired of the drama and the power and the insignificance and the responsibility and all the rest. You know you have reached a certain point in life when you welcome the weight off your shoulders. I never carried it as well as many others, but I carried it all the same. Now I can feel it leaving me.

Those mornings without residents—none of the attendings wants them. You see the patterns better than before, but there are procedures you haven't done in years, and your hands grow rusty with the passage of time.

Events, of course, don't care.

Another set of eyes comforts you. You feel less alone.

"I'll stay also," I said. "Think of me as your resident."

He laughed, in his easy way.

• • •

The man was sitting up straight with his shirt off, so pale he looked translucent, as he gasped into the plastic mask.

His belly was the color of flour. It was full and round and plump. In the upper right corner, over the liver, was a single gunshot wound, like a little red poppy.

An old man, a man whose lungs were gone, because you could see that, too, in the way he panted, and kept panting.

"Why is he sitting up?" Ben asked.

"He can't lie flat," the paramedic replied. "His spine is fused."

The nurses started working on the IVs, on both sides.

"I don't want a goddamn thing," he said, shrugging them off. "I'm done."

"It's self-inflicted," the paramedic said, like an apology. "Just so you know."

By then we all understood that he was going to die. The bullet wound over the liver. His belly swelling up before us. His terrible pallor. The expression on his face. The gasping. His age, and his frailty. The numbers on the screen.

"Give him two units," Ben said, his voice rising a little. "And let's get the level one ready. Where's trauma?"

No one answered.

"Let's get more lines," Ben said then, more steadily. Then he glanced at me and shook his head, quickly, and I knew what he meant.

The man's spine was a single column of calcium. His vertebrae had fused together over the decades as the arthritis did its work. His spine was a misshapen root, a bent chain rusted together. He couldn't lift his head off his chest, or look left and right. His shoulders were hunched and twisted. Beneath them, those murderous eyes.

"I don't want nothing!" he said, lifting his arms. "Leave me alone. Let me die."

"Give him the blood," Ben said, so we tied down his wrists and hung the uncrossed blood, icy and red and deep, against his will.

There was nothing he could do.

"I've got arthritis," he said then, looking directly at me. "Do you have arthritis?"

So I spoke to him for the first time. I asked if he wanted pain medicine.

"There's no pain medicine for arthritis," he said, and I realized that he was smiling, bitter and exalted at the same time. "There's nothing for arthritis."

I saw that he was alert. I saw that he knew exactly what he had done, and why he had done it.

"I'm not gonna put her through it anymore," he said. "She's got kids."

He was pure, and remorseless.

• • •

Then the attending trauma surgeon was there. He looked at the man for a few moments without speaking.

He's an odd figure, the surgeon, polite in person, dismissive behind your back. He seems attentive and cheerful, but he isn't, not really. In fact he's far away, and is getting tired. He comes in and out; present, absent, present again.

He's from the past, and uses the judgment of the past, which was often superior to the judgment of the present. He distrusts the tests and looks beyond them. He's always calm, as if no amount of death can touch him.

His home life is a blank space. Without surgery, he would have no natural power; surgery, for men like him, is freedom.

But he was fully present then. It was impossible not to be.

"I'm not operating on him," he said, because he is an excellent surgeon.

• • •

For once, the story was important.

He was an old man on oxygen. He had terrible rheumatoid arthritis. His hands were contorted and misshapen. He was in pain all the time. He lived with his niece in her spare room. She cared for him, and her children. He was alert. He knew what he was doing.

He spoke about her a little in the next few minutes. I realized that he loved her. I could see it very clearly.

I am trying to remember his exact words, I'm trying very hard, but they resist me. So I'm making do with the essence of his words: *This life is shit. I'm not going gently. I'm going out with fury, and I'll use my fury as a weapon against this world.*

He had a revolver in his room. Somehow, he'd managed to point it at his belly and pull the trigger.

Why not the head? I thought then. That would have been better. Maybe it was his hands.

● ● ●

People who have tried to kill themselves do not have the right to refuse medical treatment. That was what Ben was faced with. We were giving him blood already, against his will.

So the three of us discussed it, just outside the door.

"He'll never come off the vent," the surgeon said. "He won't even make it out of the OR. I don't care if they sue me. But I do care about criminal negligence."

I had not thought of that, but it was true. It was the sort of case that could, in the worst legal hands, be distorted into monstrous and criminal indifference. They are always after the money. But none of us were indifferent. None of us were indifferent at all.

It would have been so easy to take him upstairs, to the operating room, and let his death on the table protect us. He was so old, and so frail. A child could have restrained him.

• • •

But finally the surgeon reached the family on the phone, because he, too, was resolute. He spoke very carefully, and I listened. He was going to die no matter what we did, he said. The right thing was to keep him comfortable.

He hung up the phone. He turned to us, and betrayed nothing.

"All right," he said. "The family agrees. I'll admit him, but I don't think he'll last long."

Then he simply walked away, and the rest of the trauma team followed.

Ben and I looked at each other.

"I guess we should stop the blood," Ben said, and shrugged.

So we stepped back into the room and stopped the blood.

Watching someone die slowly of a gunshot wound to the belly is not something the world should ask of you.

It was a very particular kind of silence, and it took about as long as I expected. It took about twenty-five minutes.

He was entirely alert, and entirely aware, for the first fifteen minutes. His belly grew bigger before our eyes.

Still he stared, in his thunderous way, but he stopped talking. He was like a woman in labor, I thought suddenly. The belly, the determination, the gasping. It is a thought that I would like to take back, and cannot.

Gunshot wounds to the belly are cold sweat and restlessness more than pain. The pain alone is vague, bearable, imperfect. People don't scream. Usually, they are quiet.

We didn't leave him. We stayed with him. We gave him morphine, and Zofran for nausea, as the time passed, and the charts piled up in the rack behind us, and his eyes lost their fiery intention and began wandering as he gasped on.

Then he slipped away.

•   •   •

Later, we were in the hall. Ben was crying a little. He wiped his tears away with the back of his hand.

It surprised me to see his tears. I knew that I was awestruck, and shaken. I knew that I was watching everything from very far away. But I wasn't close to crying.

I can't remember exactly what words Ben said to me in the hall. But I remember their essence: *I was not prepared for that. Nothing in the textbooks, nothing I'd done before, helped me at all.*

"It was the right thing," I said. "And you handled it well."

I could have talked about pain, how that man's life was pain, and reduction, how he'd left with righteous fury in his heart and the dignity that comes with it. I could have said that the surgeon's choice also required something very like that strength. But I didn't say those things.

"It's a good sign," I said, instead.

"What do you mean?"

"It's a good sign that you're upset," I said, because the message is learned too well: *Control yourself. Reveal nothing. Let no amount of death or suffering touch you.*

Ben's tears, in the hall, reassured me. They were gone an instant after they appeared. But he didn't deny them at all. He acknowledged them.

So we kept working.

• • •

I've seen enough people die that many are wiped clean from my memory. Others are vague, imperfectly formed, in that way we half glimpse the past and reconstruct it, sifting and weighing what to keep, and what to leave behind, by mysterious means. We don't know what will stay with us and what will go, what conversations we will remember and what we will forget, what we once said and what we once heard. I suppose we choose our lessons as well, those stories of this world we want to keep, and those we want to throw away.

I would choose to remember Ben's tears. I would choose to remember the video he sent me also, a few months later, after he'd left and moved back home, of himself high in the backcountry, in the silence of the Montana wilderness in winter, skiing down through the powder in the trees, one glorious linked turn after another.

But without that man, I'm not certain I'd remember those things. I think they, too, might pass.

He was dead on the gurney. But his body was held upright by his spine.

He was still sitting up. The oxygen mask was still on his face. His eyes were half open. His mouth gaped slightly. His head was bowed, as it was in life. I was standing right beside him, that seated figure, and for an instant we were alone together.

Then I turned away, and went out to find Ben in the hall.

THE GUN SHOW

HE WAS IN HIS TWENTIES, IN A UNIFORM, THE DRIVER OF AN armored van. He'd been ambushed outside of town, in the desert, by men with rifles. They'd fled before they got the money.

The top of his skull was gone. His brain had a delicate pink hue, and a strange beauty to it, like a jellyfish unfolding in the water.

We learned the story from the police.

• • •

She was seventeen, getting off the school bus, and caught in the crossfire between two rival gang members. At first we could not find the wound, because it was a small-caliber handgun, but she lay there dead on the gurney, anyway, and finally we rolled her and saw it—a tiny blue hole, the size of a pencil eraser, high on the nape of her neck, covered by her long dark hair.

We learned the story from the paramedics.

• • •

He died quickly, and he was unlucky, because it was only a .22, and it hit him in the right side, right above the belt line, and went through his aorta. He had been running for his life up the ramps of a parking garage. His assailant shot him through the gap from the level below.

We learned the story from the police.

• • •

They were sisters, nine and ten, lying beside each other on adjacent gurneys in the trauma room.

Someone had mistaken their house for another, and fired blindly from a car with an assault rifle. The girls were sleeping together in the same bed upstairs.

My girl had a perfect red crease through her earlobe from the bullet, but that was all. She lay quietly and looked up at me, and didn't cry.

Her sister beside her, on the other side of the curtain, was dead.

We learned the story from the papers.

• • •

He was in his thirties, and came by helicopter, and was still alive. Someone had called an ambulance from the house in a little town on the Colorado border. He had two small-caliber holes in his forehead, placed very closely together, wounds that radiated experience and lethal calculation.

Meth, they said. A professional killing. We learned the story from the flight crew.

● ● ●

He was in his teens from the reservation, shot six times in the head with a .22 rifle, and everyone was drunk, and no one really knew what the story was, but the flight crew said it was a domestic dispute. He was alive, but his face was swollen and bruised and distorted, his lips puffy and his pink tongue protruding a little around the endotracheal tube.

● ● ●

She was clutching her belly and fading in and out, her dark skin pasty and gray, and you see that only toward the end. The wound was low, beneath her belly button.

The surgeons saved her by a whisker. The police didn't know the story—only that she was in a car, in a parking lot, with the motor running. Someone had seen a man near the car.

● ● ●

His son frightened him, and he was trying to get him out of the house, and when he told him he had to go, the boy fired a single shot from a pistol into his chest.

He came in almost alive, still moving a little, and what I remember most was his startled, staring blue eyes and the way he kept opening and closing his mouth and flicking out his tongue like a snake. I haven't seen this before or since.

They opened his chest in the trauma room, but it was too late. We learned the story from the police.

•   •   •

He was drunk, and harmless, but he tried to get into the wrong house, the next house over, because the houses look alike there. He kept banging on the windows and then he started kicking the front door. The father met him in the threshold with a shotgun.

Twelve gauge, right through the chest, just below the heart and just above the belly, and I thought he was dead for sure, that he had no chance with a wound like that, from that close, with that kind of weapon, but he made it to the OR alive and defied us all and somehow the surgeons saved him. No charges were filed.

We learned the story from the police.

•   •   •

He was a cop alone at night, and he pulled over the car when the plate lit up as stolen, and when he approached, the man opened the door and came out firing as he ran into the darkness. He was hit twice in the belly and once in the hip. The vest saved him, but his hip was shattered. He lay there on the gurney and stared furiously at the ceiling, tight-lipped and disciplined, crying out only when we moved him.

Later, I walked out into the hall to a sea of blue uniforms.

We learned the story from the police.

•   •   •

They were kids in a party house all night, and he owed another boy sixty dollars and told him that he didn't have it. The boy shot him through the chest with a 9 mm handgun. Then everyone was awake, and panicking, and they didn't know what to do.

They carried him outside, to the park, and left him beneath a tree, and though it was early in the morning, someone saw him under the tree and called an ambulance.

I never found out how long he was beneath the tree.

We learned the story from the papers.

•   •   •

She was an old woman, with a sign around her neck with her name and phone number in case she wandered, and her husband had shot her through the temple before killing himself because, as we learned later, he was sick also and could no longer care for her.

She lay there, blinded by the wound, with the sign, muttering incoherently, with frail papery skin, and perhaps she weighed eighty-five pounds.

I can't remember how we learned the story because it was a long time ago. But I remember her because of the sign.

● ● ●

He lay beneath me, a young man in his thirties, in office clothes, and looked directly up at me as I stood over him in my gown and mask. He told me his wife's cell phone number, that he was allergic to penicillin, and that he was going to die. He said all three things calmly, in the same tone.

He'd gone to work on an ordinary day. Another man had come into the offices, firing indiscriminately before shooting himself as the sirens grew nearer.

It was breaking news, and even then the story was on TV, but I didn't know any of that.

I told him we'd take care of him. I told him we would save him, because I always do. But he was right. When they say that there, in the trauma room, in that tone, they are speaking with an animal's authority.

We learned the story from the police.

A VISITOR

WHEN I WALKED IN TO TALK TO HER, SHE WAS SITTING FERO-
ciously on the bed with her arms crossed. It was difficult to
believe what she'd done.

"This is fucking bullshit," she said, and then got up and
started pacing around the room.

She was young and white and heavy. She had light
brown hair and blue eyes, and she looked straight at you for
too long.

"I have rights," she said.

"It's OK," I said, like someone might soothe a child.

"It's not OK!" she said, her voice rising.

"Where are you from?" I asked.

"I'm from Oklahoma!" she shrieked, startling me.

"I'll come back," I said, and retreated through the glass
door. The security guards watched.

"She needs more Ativan," her nurse said.

"Do you want us to restrain her again?" the guard asked.

"Let me think," I said as she sat down on the bed again and began to rock.

They shook their heads.

●   ●   ●

She had been there for thirty hours. A day earlier, when she was screaming and wild in the bus station, someone had called an ambulance. The paramedics had restrained her, and brought her in. She'd been sedated with Ativan and Haldol—old drugs, with decades of history behind them.

She'd slept. When she woke again, she was calmer and asked to leave. She'd made promises. But she'd also complained that her chest was hurting.

Someone had ordered a chest X-ray.

"Medicine has admitted her to the floor," the resident had said, on rounds, a few minutes before I walked into her room.

"Medicine?" I asked. "The floor?"

He shrugged.

"No one else would take her," he said.

There was a long silence. She would not be watched closely enough on the floor.

She'd stuck a needle into her heart.

●   ●   ●

People like her are unreachable, and beyond us. The language of reason might as well be birdsong in the trees.

So I sat at the desk, looking at the X-rays on the computer screen, and the CT scan and the echocardiogram, reading the notes of different medical specialties— cardiology, cardiothoracic surgery, psychiatry, and internal medicine. Each was a tribal document, arguing that responsibility rested with another of the tribes.

We had not ever seen her before. She was a visitor, passing through. We didn't know where she'd come from, or where she was going.

By then we knew that her abdomen was full of needles as well. They lit up on the X-ray like shrapnel from a forgotten war.

The needles were all the same, white among the shadows on the screen. They were three-inch sewing needles, exactly like those my grandmother had used with a thimble. Looking at the screen, they seemed as if they'd been chosen with care, in their identical natures, as if she were following a form of order, repeating a ritual over and over again. They must have been there for a long time, because her belly was pale and unmarked.

The body can endure sewing needles in the abdomen. The odds favor you; the chances of damaging the bowel or a major blood vessel are low.

But finally—no one knew exactly when—she'd found the perfect place, just below her left breast and between her ribs.

She'd buried the needle into her chest until it disappeared, through the tough fibrous tissue of the pericardium and deep into the cardiac muscle itself. It takes force to do this. The point of the needle extended into the chamber of the ventricle and remained there.

If you looked closely, you could see a tiny, bloodless pinprick. But that was all.

The body can't endure a sewing needle in the heart for very long. At any time she might have started bleeding in earnest, as her heart whipped the needle back and forth like a conductor's baton.

A brutal operation awaited her. But for thirty hours, she'd refused to let the surgeons touch her.

"We need to take the needle out," I said, back in the room a few minutes later, trying to reason with her yet again, as if we were speaking about the ordinary world. "That's why your chest is hurting."

Again the long, clear stare.

"I put the needle in because my chest was hurting," she said, patiently. "You can't take it out. I need it."

"Do you want to kill yourself?"

"Of course not," she said.

She looked perfectly well.

● ● ●

In the past, there were endings. There were mental institutions. The institutions were vast, and I rotated through one as a medical student. From a distance, it looked like the campus of a liberal arts college, with red brick buildings and sheets of mown green grass and oaks.

Mental institutions were not indecent places. Depravity existed within them, but so did responsibility and kindness.

But then the funding was cut, and great numbers of patients were either cast out into the street or placed in poorly regulated group homes. It's a story that has been told many times, to a collective shrug. Money for innocence is one thing. Money for the unreachable and the frightening is another.

No one likes the mentally ill. They repel us because they are so close to us.

I called the surgeon.

• • •

"I spent forty-five minutes with that woman this afternoon," he said. "What do you want me to do? Tie her up and operate on her without her consent?"

"That's what psychiatry wrote in the chart."

"I can't do that. She has to cooperate afterwards. What is she going to do to the wound?"

I thought about it—an open wound on her chest, with her heart beating an inch or two beneath it.

"I won't admit her," he said, to settle the matter, "if she won't consent."

So I made a dozen more calls that night. I spoke to the hospital administrator, to other surgeons, and psychiatrists, and the internal medicine doctors, and the ICU. At first I enjoyed the absurdity of it. It felt surreal, compelling, and pleasantly, seductively righteous. But after a while, I got tired of my righteousness, tired of the same heavy, obvious arguments and the quicksand of pages going unanswered. I got nowhere because no one wanted her, and I understood this because I didn't want her, either. I wanted her gone.

Somewhere between phone calls it occurred to me that no one in this story was rational. The needle was invisible to the naked eye, and therefore abstract. Abstract knowledge is as powerless as abstract pity. It so rarely moves us to act. We know better, and yet we follow our instincts anyway. The needle in her heart seemed like a statistic, a graph of rising temperatures instead of heat.

. . .

Finally she was wheeled upstairs to the medicine floor, tied to the gurney with leather straps, my responsibility no longer.

Usually, when I go home, I don't look back. I don't think about those I've seen; I don't have dreams. I leave them behind.

But for the next few days I followed her anyway, on the computer, from a distance. Partly I followed her from curiosity, and partly I followed her from frustration, but mostly I followed her because she revealed so much. She spoke so clearly to our judgments, to our values and decisions and choice of responsibilities, to our primal natures, to how often the rational intellect simply launders our animal selves, offering desires as arguments.

The days passed. Still no one wanted her. The debates went on; the meetings were held; the ethicists appeared, moving without urgency toward the inevitable as her heart began to show increasing signs of damage. Then, finally, on the third day, the needle began to move.

It was the movement that did it. This was knowledge that could be felt. When the needle started moving, they took her to the operating room.

•   •   •

I read the surgeon's note from my couch at home, many miles away.

They wheeled her into the room. They put IVs in her arms. They anesthetized her, and intubated her, and put her on the ventilator.

They prepped her chest, scrubbing it with brown Betadine, in their gowns and gloves and masks. Then they draped her, with blue surgical sheets, until only her chest was visible. I could picture it all.

They made a vertical incision down the center of her chest, touching the cautery to the blood vessels, which crackle and release little wisps of smoke into the air.

They ran a saw up her sternum, wiping bonemeal and blood from the blade. They spread her chest apart and exposed her beating heart for the first time. They could see the needle then.

They put her on the heart/lung bypass machine, which diverts the blood flow to her heart. They cooled her heart with saline, and then they stopped it with potassium.

A cold, still heart, with the body alive around it, is hard to imagine. But it is routine now, in every city in the developed world, the product of the rational mind, and so many centuries of inquiry.

They made a tiny incision in the ventricle. They felt for the needle with gloved fingers, and then they pulled it out.

They sutured the incisions tight. They closed the needle's pinprick in the ventricle. They washed her heart.

Then they released scarlet, oxygenated blood from the bypass machine back into it. They waited, and watched, as her heart slowly grew red and warm.

It's astonishing to watch a heart spring to life again. I saw it many years ago, as a student, but it is not something you forget.

They took her off the pump. They put in the drains, and they sewed her sternum shut with rust-proof wires that will remain in her future casket for a thousand years. Finally they closed the wound—a foot-long gash, down the center of her chest.

All of it, from beginning to end, took less than two hours.

They wheeled her to the ICU.

● ● ●

The bodies of the young come back fast. A few hours later she was off the ventilator, and two days after that she was walking in the hall. During the week she spent in the hospital, she was watched all the time, by a sitter, who sat beside her like a friend.

On the eighth day, she was discharged to the mental health center, under the care of the psychiatrists.

The beds there are few, and valuable, and the line for them is long. There are always people coming, full of voices and visions, people who are dangers to themselves or to others, people who won't take their medications, whose families have abandoned them, who have no money or insurance, people with one foot in this world and one in another.

The psychiatrists need to move them through. They have no choice. And so they seize on anything—a promise, an apology. A recognition. The claim of fainter voices, or the desire to live.

The note described her as cooperative. She was feeling better. She did not want to put needles in herself anymore. She wanted to leave, and continue on to California. She had friends there, someone she would meet. She still had her ticket. She would take her medicine. She was not a leopard in a cage, pacing around and around again.

They got her a cab to the bus station.

THE GESTURE

THE NURSE CAME TO US. A MAN, WAITING IN THE CUBICLE to be seen, had swallowed pills from a bottle in his backpack, and then told the nurse what he had done, proudly, like he'd won a prize.

"What did he take?" I asked, and sighed, getting up, grabbing his chart, and making my way toward the room, because people who do this are exhausting. I've seen so many of them.

"I'm not sure," the nurse replied. "It wasn't marked. I took his backpack away from him."

"It just happened?"

"Yes," she said. "He's an idiot."

I walked into the room, and he was sitting on the gurney. He was in his early forties, a bit heavy, with dyed-blond hair, but I would not recognize him now. I can't picture his face.

"Sir," I said. "What did you take?"

"I don't know," he replied. "It was nothing."

"Do you have the pill bottle?" I asked the nurse.

"I'll get it," she said, and left the room.

I must have looked at his chart. I know he was there for an unrelated complaint, but I can't remember what it was—a headache, abdominal pain, something. I do remember that he wore mascara, and that later it ran. But that's all.

"Why did you do that?" I asked him.

"I don't know," he said. "My boyfriend is an asshole."

"Did you have a fight with him?"

"He's in the lobby," he said after a moment.

There was a cell phone on the sheet.

The nurse came back and handed me a large plastic pill bottle. It was oversized, the kind used by people who take medication every day and don't want the inconvenience of refills.

The bottle was about half full of small green pills. But the prescription label had been peeled off.

Pills are marked with codes. I shook one out on my palm, and looked at it, and saw the numbers.

"We're going to find out what these are," I said. "So just tell me."

"Fine," he replied, as if I had inconvenienced him. "It's Inderal. It's for my blood pressure."

"Inderal? How many did you take?"

Because suddenly this story had changed. Inderal is dangerous.

"I took about half," he said. "I'm sorry, I didn't mean it."

"You took half the bottle? When?"

He was wringing his hands.

"I don't know."

The nurse looked at me.

"They were in his backpack," she said. "He just told me he'd done it. I think it was right after his boyfriend called. That was about ten minutes ago."

The bottle had space for hundreds of pills.

• • •

Pumping the stomach is a thing of the past. It is hardly done anymore, and only in specific circumstances. But this was one of those circumstances.

So we rolled him out of the cubicle and down the long hall to the resuscitation room. By then, the resident was beside me.

We were curt with him, and we were clear. We were going to pump his stomach. He could cooperate and help us, or he could resist. If he resisted, we were going to do it anyway. It would be better for everyone if he helped us.

I was angry with him. He was consuming our time and our money and our fears. He was too old to behave in this way. It was the gesture of an adolescent, not the act of a grown man. And it was dangerous, far more dangerous than he understood. He thought that he was safe. He thought that he was in a hospital, surrounded by doctors and lights.

He gave us no choice but to speak to him like a child: *If you fight us, we will tie you down.*

And then the unspoken thoughts: *You have frightened me with your stupidity, and you don't even want to die.*

But he didn't fight us. By then he was fully remorseful.

"I'm sorry, I didn't mean it," he said, and began to cry.

• • •

We used a tube known as an Ewald. It is almost the diameter of a garden hose. You put a bite block in the mouth, insert the tube through the hole in the bite block, advance it into the stomach, pour warm water down the tube, and then attach the tube to suction. It is not pleasant to endure.

But he sat still, and opened his mouth for the bite block, and bent his head when we told him to, and swallowed when we told him to, even as he gagged and choked during the insertion, because it is impossible not to. He understood the severity by then, and I remember how he kept apologizing, how he kept trying to take it back.

We got the tube into his stomach, and taped it to his nose as he breathed around it, and then we started the lavage.

There were pill fragments in a green sludge, lots of them, and so I knew he was telling the truth. He had done what he said he had done.

A minute or two later, it hit him like a train.

He went gray and limp. One moment he'd been gagging on the tube, and the next he was falling against us, his head gone loose, like someone nodding off to sleep.

"Take it out," I said, and the tech grabbed the Ewald and pulled it free of his mouth, trailing threads of mucus, and then we laid him flat, and it all began.

• • •

Drugs like Inderal slow the heart rate and lower blood pressure. In high doses, they do what they did to him. They are ferocious and lethal. He didn't know this, because in the eyes of the public, one little pill is very like another.

He had no pulse. His heart was not beating. He was not breathing. Just like that, he was gone.

Most medical codes end in death. Most codes are for the elderly, whose hearts have stopped, whose lungs have finally given out, whose aneurysms have burst. An ambulance is called, and the ritual of the declarative end unfolds. You have an algorithm to follow, and you follow it. After a while, if they don't come back, you let them go, and you forget them.

But I thought we might save him. I thought we had a chance.

So we gave him everything we could think of and we kept going. We gave him epinephrine, and atropine, and enormous doses of insulin, and glucose, and glucagon, and vasopressin, and calcium, and intralipid, and we intubated him, and we put in a central line, and we tried to drive his heart with the external pacer, and gave him more epinephrine, and more glucagon, and bicarbonate, and then the toxicologist was there, in his suit, looking at the strip and asking the questions. One by one the techs and the nurses and the residents took turns doing CPR. You get tired soon because you have to use force to do it well. You have to lock your elbows and bear down and try to drive the sternum to the spine, and not worry if you feel ribs popping after a while.

"I don't think there's anything else to do," the toxicologist said. "I think you've done everything."

But we kept coding him anyway, because he was still young, and he'd taken the pills in a hospital, surrounded by doctors and lights, and had announced it like news.

THE STORY OF THE
CHERRY PICKER

I DON'T KNOW IF HE OWNED THE CHERRY PICKER OR IF IT belonged to a company. I know only that he kept it parked outside his house on a flatbed trailer.

It was a suburban neighborhood, with wide streets and brown houses, the desert on one side, the Sandia Mountains on the other. The Sandias leap up, blue and silent, with their gray granite cliffs and ponderosa pine, and you can see them from a hundred miles on a good day.

A cherry picker is a little like a Ferris wheel. It goes up, and gives you a view, and comes down. It must be exhilarating at times, on the roadside, standing in the bucket as you rise toward the job, and the expanse of the city and the desert and the mountains are all around you.

Thirty feet is not that high. You can survive a fall of thirty feet.

Sometimes, he would take the kids in the neighborhood

up in the air on the cherry picker. They'd stand in the bucket and rise. He'd be with them, working the levers.

I learned this later. It should have been fine.

* * *

The cries of children trouble me, and the stakes are so great. There are doses to remember, and different tests to order. When they are very young and cannot tell you what's wrong, they are like small quivering animals.

Sometimes, you hurt them. You have to draw blood and start IVs, and their veins are small and hard to find. You have to suture lacerations as they writhe.

They are strong also. When they resist, it is difficult to hold a small child down, even in infancy. They fight you, and shriek, as you look in their ears. Sometimes you have to sedate them for a scan or a procedure, and it's always a little dangerous, and you have to watch so closely.

Children are tough and delicate at the same time. Their hearts can take a very great deal. They can bleed down to almost nothing and come back. But when they go, they go quickly. Their mouths are small. Their vocal cords are tiny glittering things. Sometimes the tubes you use are hardly bigger than a straw. The stakes are so great.

Sickness in children is different from sickness in adults. It is more blinding. They have no agency. It is never their fault. When children fall thirty feet, it was because they didn't understand, or were taken there, or climbed, because children like to climb. They climb because there was once safety in the trees.

Original sin—I think about it. But you cannot tell. You have to give them time. If there is cruelty within them, then so, too, there is goodness. The promise of darkness is also the promise of light. This is what we mean by innocence.

A big mistake with a child is an existential threat.

I stopped seeing children because my hospital is large, and there is a dedicated emergency room for them. It was easy to avoid them, in the unconscious way that we make choices.

But that day, when the cherry picker fell, it wasn't a thirty-foot fall. The cherry picker was a lever arm, and it weighed ten thousand pounds. When it fell, the basket struck the ground like an enormous whip. I know this because I read about cherry pickers afterward. It was far more than a thirty-foot fall.

● ● ●

Too many dying children came in all at once. The trauma team was overwhelmed. Pages went out everywhere, but we had no warning. So every doctor in the ED got a child.

My child was desperately injured. My child lay there intubated, with a chest tube, and still the oxygen saturation would not rise.

I stood there, and I could not understand why. The

chest X-ray looked OK. The tube seemed in. The blood was hanging, in the right amount. But the oxygen saturation would not rise. We checked, and checked again, and it was the same.

I looked at the child. I cannot remember if it was a little girl or a little boy. I understand what this reveals. I felt only my own helplessness as we waited, and I went through it in my mind again. I got another chest X-ray. It looked the same. I checked the tube again. I checked the probe again. The blood was running in the lines. The CT scanner was waiting. The oxygen saturation would not rise.

The children were quiet. When people are hurt very badly, when they are dying in earnest, usually they cannot cry out.

None of the children cried out. But one by one they died, either there, in the ER, or upstairs, in the OR or in the ICU. By morning, all of them were gone.

And the man, the friendly neighbor who had taken them up—he was gone as well. He was dead on the ground beside his cherry picker. For that, I suspect, he would have been grateful.

He'd raised the cherry picker from the trailer rather than bracing it as he should have done. In still air, nothing would have happened. They would have gone up and come back down, an easy rise and fall. They'd done it before. They'd look out, from the basket, and then descend again to dinner, because it was after school.

A gust of wind had come. Just that—a gust. Some-

times they are powerful during that time of year. Sometimes they kick up dust devils over the sage.

When the gust caught them and the cherry picker
started to tip, they all clung to the basket rather than
jumping, because that is what people do.

•   •   •

Order—our love for it is great. I thought about the gust
of wind, and how to the wind, the cherry picker had no
significance at all. We think of the dangers to our children as men in vans, as swimming pools and toddlers, as
unbuckled seat belts and fire, or pills in cabinets, but this
danger came from kindness. A decent guy, and the kids
pestering him—*take us up, take us up*—after school. He
was probably tired at the end of the day.

The cherry picker changed me, a little. I'm not sure
why, because these are lessons that I've learned a thousand
times. I've been taught them again and again. Perhaps it is
because I didn't act well. I wasn't calm enough, because it
was a child. Because the world had shifted and distorted
all at once. I know this most of all because my memory of
the event is a little unclear. It lacks radiance. It is a duller
memory, a little muddy, a little hazy, and its opacity is
ominous. Of this, I am certain.

I can't think clearly about the cherry picker. And
thinking clearly is comfort. Thinking clearly is a balm and
a solace. First the beginning, then the middle, and then the
end. Tell any story with beauty and order, and perhaps the
world will follow.

But the story of the cherry picker is not a story at all. It's a recitation. The basket rose, and the wind blew.

The wind blew. The cherry picker swayed back and forth. But he lowered the basket quickly to the ground. He opened the door. One by one the kids jumped out.

Everyone went home.

THE HORSE

FROM THE NECK UP, CATHERINE LOOKED LIKE AN ORDINARY woman, mild, scholarly, with thin wire glasses and bright eyes behind them.

From the neck down she was hardly a woman at all. Her body was enormous, so full of fluid that she could barely move. We had to lift her from the paramedics' gurney to the bed as she gasped into the oxygen mask.

Her shirt was off, exposing her breasts until we drew up the sheet. Her skin was gray and damp. Her hands and feet were blue. She felt cold to the touch, like a bag of flour in the refrigerator.

"I can't breathe, I'm so nauseated, help me."

The words came one by one, between breaths, like beads on a string.

"Help me," she said, again, and then she began to cry. She cried like she gasped, without restraint.

I asked her if she wanted to be on a ventilator. I asked

her what she'd like us to do if her heart stopped. I had to get straight to it.

"I don't want any of that," she said. "I'm so nauseated. I'm so afraid. Please help me."

"What would you like me to do?"

"Something for nausea. Something for pain. Oh my God."

So she absolved me, because she had the courage to give up. I would only have to watch.

She was terrifying.

• • •

Primary pulmonary hypertension is mysterious and deadly and rare. The immune system attacks the pulmonary blood vessels. The arteries become stiff and inflamed. The heart struggles to pump blood through them, and in a few years it gives out. There's no cure. The failure begins with breathlessness, and it ends like it did for her. She'd called the ambulance from home. The nurses came only a few times a week, and they weren't there.

She should have been in a hospice. Instead, she was alone in her house.

It comforts us to name things. It comforts us especially when a disease is rare, when the odds are long, when the chances are great that it will never touch us. Something else, you think, but not this, and it is an uncertainty you embrace. Her suffering was safely hers. It had chosen her, like a celebrity, like a single star in the sky.

• • •

The hospital was full. The hospital is always full.

There was no place to put her upstairs, no place for hours. Just the ER, in the afternoon, with a crowd waiting in the lobby.

For a little while, she stopped crying.

"Is there anyone you'd like us to call?"

"My mother is elderly. She doesn't drive. Please don't call her."

So we didn't call her. But she wanted to talk, and she didn't want to be alone, and we could see it.

I asked her what she did and where she was from, as if we were getting to know each other.

"I'm an artist," she said, her lips forming little circles, like rings in a pond.

"What kind of artist?"

"I paint watercolors."

Then she began sobbing again, opening and closing her hands.

So it was up to me—how much morphine to give her?

• • •

Once, when I was a child, I rode in a horse-drawn cart for a few miles down the coast of India. We were going to a hotel outside of town.

We'd gotten into the cart in the dark, on a soft, hot, windy night, and set off briskly. The horse's hooves knocked on the dirt road, and the surf roared in the distance past the beaches. The moon was out among the clouds, and it was shadowy and exciting.

But after a while it became clear how weak the horse was. It could not sustain the pace. We could barely see it in the dark. The driver whipped it, and whipped it again as it flagged, trying to get back to town a little faster and pick up another fare. I remember grabbing his arm and asking him to stop. He looked at me, puzzled and annoyed—as if to say, *It's just a horse. Who are you?*

● ● ●

Morphine—the question weighed upon me. Too much would kill her, not enough was worse. The line was so thin. She was dying, but I still didn't want to kill her. It's a primal thing, that decision. You want to do something. The desire is very powerful. You have to resist yourself: doing nothing is a greater form of discipline.

But that next step—pushing her over, and being done with it—is an even harder act. It requires a particular kind of courage. You can't be wrong. You can have no doubt.

I was working, and there were many other patients to see. I spent only a few minutes on her. There were new hands to shake, tests to order, there were residents and students, there were EKGs to read, there were notes to type and questions to answer, there were ambulances coming and people walking by, and through it all the cell phone on my hip kept ringing.

I wanted to turn away, and I did. But I felt her presence nonetheless as the hours passed. I knew she was behind the curtain in the corner, crying and pleading. Her begging was general, like that offered to a torturer, as the nurse sat with her and held her hand.

So I compromised. I gave her almost enough.

• • •

It took hours for her terrible alertness to fade. But finally, inevitably, she became confused.

And so she became less terrifying. The gulf between the watched and the watchers grew greater. Her breathing slowed, her head started to fall back, her mouth began to gape. She'd gasp awake again, look up, blink a little. She moaned like someone who is dreaming in their sleep.

The nurse still sat with her, and still held her hand, but she had to get up and work also. She got the worst of it. A young woman, not yet thirty, and her life isn't easy; she has bills and children and troubles. I know this because the nurses talk to one another, and I hear it. They talk to me also, sometimes.

By then we'd turned off the monitor so the alarms didn't ring.

The task had fallen to her: it was she who sat with her when she could, and looked straight at her, and tried to comfort her as the hours passed and the bright lights fell.

• • •

The end of a person's life occupies, at most, a single room. But it fills that room, and there is a sense of reverence within it. Everyone can feel it.

I felt it. But not as much as I would have when I was young, and that's something that's stayed with me also. If you're unlucky, the coldness of your own heart gathers strength as time passes, and forces a certain clarity upon you. You become part of the indifferences. Experience makes you stop seeing the cruelties of the world for what they are. It's something that you must resist, and something that you must remind yourself to resist.

Finally, she stopped gasping herself awake. Her head fell back for good, and her mouth opened behind the mask. Her glasses were still on. Her breaths became further and further apart. Her face slowly went blue, until it matched the color of her hands. I was there for enough of it.

A watercolorist—the mildest, the most harmless of women.

● ● ●

The hotel was a concrete building, and the nicest place for miles. It was lit up in floodlights, and there were floodlights down by the water, too, illuminating the surf for a nighttime swim. It seemed miraculous, after the third-class trains, and the heat, and the beggars everywhere.

We got out of the cart, with its faded tassels, its red and blue peeling paint, its worn cushions in the back. The bellboys converged for the bags, and we got a good look at the horse for the first time.

It stood there under the lights, and I remember how deeply it shocked me, with its ribs like the branches of a tree, and its mousy coat worn hairless where the yoke rubbed.

I would like to believe that I would still see the horse. After all these years, I would like to think that I would not look through it entirely.

The driver clucked and shook the reins, and they moved off into the dark again.

•   •   •

I pronounced her dead, which is a little ritual from the past. I put my stethoscope on her chest and pretended to listen. I shone the light in her eyes. Then I picked a time, and the nurse wrote it down.

It was false precision—3:32 P.M., not 3:30. You don't round up, or down. You choose a number that looks measured instead of guessed.

We covered her body, and moved it to the decontamination room to wait for the funeral home. The housekeeper came, and made the bed again, and wiped the floor. A new patient was wheeled in, someone with no idea of what had just happened there.

Then the nurse who had sat with her, and tried her best for her, stepped out into the hall. I followed her, and touched her shoulder, and said something to her—I forget what exactly—but she shrugged me off and walked away.

Ten minutes later she was back, dark-eyed, quiet. I knew she'd been crying, gathering herself again, and as I looked at her, I thought, *She's so young. She has so much left to see.*

THE SUNFLOWER

SHE LAY THERE LISTENING, EXPRESSIONLESS, WITH HER DAUGHters beside her. The Vietnamese translator spoke through a video console from California.

"She says she's had the pain for a year."

A tiny woman, thin as a child. Her daughters looked at me flatly, and then the eldest turned and addressed the translator through the machine.

"She is saying that if you don't know what it is, then why is she getting worse?"

"I don't know," I replied. "Ask her why she missed her appointment with the specialist."

The translator spoke at length, and I wondered what else was being said, or why.

"She didn't have a ride," the translator said finally.

"Ask her if she is lonely," I said, looking at her daughters. "I know she lives by herself."

The translator spoke, and the woman answered.

"No," the translator said. "She says she is not lonely. She says she has a pain in her stomach."

"Let me see if the labs are back," I said, because it was an endless circle.

They exchanged their looks again.

Back at the desk, I opened the note on the computer and looked at the dozens of visits, stretching back for years, and it was then that I saw his name.

Joe.

I suddenly understood that he had been sitting where I was sitting, typing his note into the same computer, about the same woman and the same daughters, just a little way back in the past. It felt entirely ghostly to read what he'd written about that woman, and to see how exactly his thoughts about her echoed my own.

•   •   •

Joe was an enormous man. His weight hung over him like a question no one wanted to ask. How could he allow himself to be so big? Why could he not have more discipline, more strength, more belief in himself?

Every time he came in for the long night, where he would work alone in the low-acuity section of the ER, he would meditate in the break room for a few minutes before his shift began. He would sit down and shut his eyes, with his hands outstretched, and listen to something through earbuds on his cell phone. It might have been music; it might have been a sermon, or a self-help

book, or a mantra of some kind; I never heard it because he kept the volume low. And if I entered the break
room to find him there, with his eyes closed, listening
to whatever it was, I always left, I always stepped away,
because it was unsettling to see him filling the chair in
his cowboy boots, with his black goatee and his ponytail,
and all that weight, so visibly preparing for the hours
ahead.

The beginning of a twelve-hour night shift, when you
drive in, feels a little bit like the end. You feel the weight
of exhaustion before you, because it is always an ordeal.
Everything feels a little off—your headlights swinging
through the empty parking garage, the sound of your own
footsteps in the dark as you walk through it.

Once you're working, it becomes ordinary again, the
familiar struggle against your body at three or four in
the morning, that flash of alertness when pagers go off,
the charts before you with their troubles, the way you
must remind yourself to pay attention, the way you can
feel your own mind grow duller, and then sharpen again
as the sun rises. You put aside the little whisper of fear
in the background that never leaves you, alone with the
events and the decisions they require.

Joe worked a lot of nights. He remained a stranger,
someone no one wanted to get close to. Sometimes things
were said, and they weren't kind remarks, but no one really
cared enough for outright cruelty. It was mostly indifference, a faintly dismissive tolerance, against the background

of the lobby and its heartlessness, a man whose single distinguishing feature was the size of his body.

But Joe was good at his job, which his weight obscured. His body and his quiet manner hid his private intelligence from everyone. And he worked alone, all night, steadily, without fail. I trusted him without respecting him, and that's the truth. I judged him, just like everyone else.

•   •   •

Sometimes, there are those who reveal our fears. *I could be him*, you think. But I thought of him as someone I could never be, as someone with whom I had nothing whatever in common. I didn't like him or dislike him. We both saw the other with almost perfect indifference, neither friendly nor unfriendly. In the end I understood that I didn't know him at all, where he'd come from or where he was going.

But the nurses told stories. They talked of passions, of ex-wives and girlfriends, and they spoke of a child—an infant—somewhere in the middle. The child was sick, at one point, that much I did know. They knew this because he had told them, in his lower moments, of which, I think, there were many.

Once, he talked to me about motorcycles, the Harley he rode, on the weekends, out through the desert to various towns, and he told me he liked taking photographs. Small talk, I suppose, the kind you reach back into your memory to find after things happen. He'd come from somewhere in the Midwest. He'd worked in a number of hospitals. Now he was here for a while—a few years, maybe, he wasn't sure.

I suppose the reason that I started thinking about Joe again was because he was, for me, a kind of test. Can you be merciful? Can you find compassion for those toward whom you feel nothing?

Our cold hearts reveal us.

•   •   •

Nursing is full of the echoes of damage. You can't talk about this. But so many in nursing put up with bad partners and struggling children; so many are caught in the endless cycle of pleasing the weak and appeasing the cruel. The pattern repeats, and repeats most of all in the personalities of the night, in those who choose the night because they do not feel as if they quite belong in the day. The night shift is for the uneasy and the unsettled, those on the edges of the crowd, those who are lonely, and sometimes have secrets, and want to step away, just a little, from the rest of the world.

A hospital is purer at night than in the day. There is the sense of being left alone, where no one bothers you, where it is just the work, where the quiet is greater and longer as the small hours pass. The price you pay for the slower procession, and the greater intimacies, where Joe confessed his life to the nurses around him, is that time falls upon you more heavily. You feel the ache in your body, and your mind slows, and you have to get up, and walk around, and shake yourself awake, because you always have to pay attention.

The nurses helped him, as they do.

● ● ●

Later, Anna told me what happened to Joe. Anna is a quiet woman, dark-haired, dark-eyed, born and raised in this city—a woman who hasn't traveled far, with children of her own and a life she doesn't talk about very much or share. She's been a nurse for a long time. Her family has been in New Mexico for hundreds of years.

Joe had split from his wife and moved to a cheap apartment. His wife kept the child, their infant daughter. Much of his money went to them, and it was enough but no more, because Joe was a physician's assistant, not a doctor, and so was paid half the salary for the same work. This is the future. It's business.

So people like Joe are in demand. There are jobs everywhere, and choices. But he liked the desert and the empty spaces, and he couldn't leave anyway if he wanted to help raise his little girl.

Joe was lonely, and for companionship, he got a dog at a shelter. The dog was young, and full of energy, a thin brown New Mexican mix, and I saw a picture of it once, on a cell phone, as he held it in his arms.

He was trying to lose weight. He walked his dog. The dog was his companion and his encouragement. On a snowy morning, with ice on the sidewalks, and the dog pulling hard on the leash, he slipped.

Joe weighed over three hundred pounds, and he was of average height. He fell with enough force to tear his hamstring almost completely off the bone.

Anna took the dog and cared for it, and she visited him in the hospital after the surgery to reattach the muscle, and she visited him in rehab later, as he started to get up and walk.

I didn't know any of this at the time. I only knew I hadn't seen him for a couple of weeks. I didn't even realize that he was gone.

* * *

I'm not sure if I would have talked as easily to the nurses during the day and learned the story. They were women in their thirties and forties, with jobs and troubles and families, and they knew all the details of his life—the turmoil, the loneliness—and even, at times, they gossiped and laughed about them, but in the end they helped him when he fell, because he had no one else.

It's an American story, that loneliness. It's one of the struggles of being white in America, which no one ever talks about, because white people are so often alone at the end. Their companions come in ones and twos or not at all. The white American family is scattered and indifferent and far away, taken by circumstance, by jobs and marriages and divorces and all the rest, to other places, to other states and cities, lost to individualism, to the marketplace and the idea that everyone must make their way alone.

The surgery to reattach a hamstring that has been torn off the bone is brutal and painful. And when you are that heavy, everything is worse—the weeping wound, the bedpan, the catheter, the sweat and the odor, the impossibility of keeping clean. So it was an ordeal for him, a terrible one, trying to heal with all that weight, and all that moisture in the wound.

He was in a facility, on the slow recovery, on the antibiotics, when it happened.

He texted Anna for help when they ignored him. She showed the texts to me because she'd saved them. She wasn't crying, but she was close to crying.

I think I'm having a PE, he'd written. They won't answer the call button.

I held her phone in my hand, and read the desperate, terrified texts one after the other, texts she hadn't seen for a few hours, after it was too late. I imagined him, enormous in the bed on a Sunday afternoon, unable to get up on his own, trapped by the indifference and ignorance of the attendants who didn't recognize the emergency even though he'd told them what was wrong. He must have felt like he was being buried alive.

He was correct also. It was a PE, a pulmonary embolism, a blood clot rising into his lungs, because he had been on his back for weeks. It causes pain, and breathlessness, and a sense of terror when the clot is large, as his was.

So he lay there, texting and waiting and crying out for help, and when the ambulance finally arrived, he was dead.

●   ●   ●

For so many, the story ends there—an obituary in the paper, a service by a grave, absences that are filled in again so quickly by the ordinary lives passing around them, the primal indifference we have toward almost everyone we know.

It takes effort to resist it. It takes something larger and perhaps better than the self. It takes someone like Anna, who surprised me entirely with what she did next.

She organized an exhibition of his photographs. She found them, on his computer, in his apartment, because he had given her a key.

The nurses who had worked with him for those few years contributed money to a pool. They're not poor, but bills are often a struggle. A hundred dollars means something to them. A hundred dollars is more than a gesture. But they each kicked in a hundred dollars, and then they sat down and looked through thousands of photographs.

Joe had taken it seriously. He had a good camera. And he rode out most weekends, or on days off, on his Harley, and New Mexico is full of empty roads, full of blue sky and rock formations, and piñon pine, and cedar, and sage brush, and coyotes, and elk in the northern grasslands, and black lava in the west, and miles of empty desert in the south.

They printed and mounted the ones they liked the best. They rented one of the public libraries at the edge of town. They bought white cake from a supermarket, and bottles of Coke. And they posted flyers in the ER, and sent out a group email.

● ● ●

The library was forty minutes away from my house, and I almost didn't go. The truth is that I forced myself to go, because Joe, for me, was not one of the precious few. Had our positions been reversed, I never would have expected him to come.

I was a bit older than he. He was able to carry his weight as he did because he was only in his early forties. In a few more years it would have become impossible for him.

But I did it; I got in my car and followed the directions out to the suburbs, where the houses were a little cheaper and the crime not as bad, where people with jobs but not much money can raise a family well enough to keep it all going.

The nurses were there, with their husbands—just a few, and they were surprised to see me, but they were also happy to see me as we all ate white cake with vanilla frosting and drank a little Coke.

Their budget was small, and so there weren't many photographs on the wall. Barely a dozen, in frames from Hobby Lobby, each with a careful, handwritten title.

Most of the photographs were good. They were good

in the way that photographs for sale in local restaurants are good, or the way that photographs that win prizes in high schools are good. An arch, a door, an adobe wall. Red chili peppers in the sun. An empty road. Aspens in the height of color up by Santa Fe. A man riding a horse, with the sun behind him and white clouds over the Sangres. And Carlsbad, deep in the cave underground, with floodlights.

But the sunflower was better than the rest. It was taken up close, with an excellent lens, and leapt out at me, in black and white. I stood in front of it, and let it fill me up a little, because it was beautiful.

The event was sparsely attended. The time was bad, and it was a long way, and all the rest. None of the other doctors came. I didn't blame them. The reasons for my presence were complex, but I think in the end I went because the nurses had humbled me with their kindness. I saw it so clearly, and when I measured myself against it, I knew that I fell short. A forty-minute drive on a dark evening felt like an effort, and a sacrifice, in service of self-congratulation, but I was wrong, because when I looked at the sunflower, I did, in the end, feel something for him after all. I think now that I went for that, because the absence of sadness feels so dangerous and empty.

● ● ●

His dictation read:

> [Sixty-six]-year-old Vietnamese-speaking female with
> a history of chronic abdominal pain and multiple prior

ED visits presents today with the same complaint. She is well appearing with normal laboratory values. I do not believe imaging is indicated. I have encouraged her to follow up with her primary care provider, to not miss future appointments as she has done many times in the past, and to return for worsening symptoms. I explained this in detail to the family, who are present, with the assistance of the Vietnamese interpreter.

Then he discharged her, with a diagnosis of chronic abdominal pain.

So I walked back in the room, and I called the interpreter in California again, on the video console, as we waited in silence.

"I'm sending her home," I said when the screen came to life. "All of her tests today are normal. She needs to keep her appointments. I've looked at the records and I know this has happened many times before. Do they understand?"

The interpreter spoke through the speaker, and her voice seemed very far away, disembodied and tinny, as the patient and her daughters strained to hear.

They answered, and then the interpreter spoke again.

"Yes," she said. "They understand."

THE SNOWSTORM

So he called me, because he was having pain, and was breathless, and the rest, and he wouldn't go in, and he'd chosen to live alone, to be alone with himself at the end.

I yelled at him like a good son. I threatened him as circumstances required, and insisted, because I heard it in his voice and I knew what he wanted, which was to be safe, and alive, and cared for. He made me work for it all the same, headstrong and foolish and afraid, like so many old men when their bodies start to fail them, and they act like small boys.

But that night, he listened, he went to the hospital, and they saved him. They took him to the cath lab, and he had a blockage, a big one, poised to kill him, larger than they or I had expected, a fat, sweet berry waiting to be plucked.

It's a ten-hour drive. I left when I could.

• • •

So I was driving north to Colorado in the darkness, north in the new car, with its new-car smell and its leather seats and

the stereo and its colored numbers on the dash. I knew
the pressure of my tires and the temperature outside, and
I knew exactly how far away I was.

By then I knew what they'd found also, and I knew
that he was all right, that his heart was undamaged, but as
I drove, I felt afraid anyway. I was at that time in life when
your parents begin their departures, and the generation
before you is about to melt away and leave you standing
with the crowd at your back.

It began to snow, just a little at first, but then heavy in
the lights, and so I turned them down, and driving on a
dark empty road through the mountains in a snowstorm
is like falling into the deep. I could feel the heated seats
beneath me, and the cold breath of the windshield against
my knuckles, as the road went white and the temperature
fell to three degrees above zero. I played classical music, and
I kept driving, mesmerized, through the night, thinking
the thoughts that we have at such times, those moments
where life formally marks its passage, like the slow count
of fingers—one and two and three.

• • •

The hospital that I reached in the morning was vast and
hopelessly expensive. Hospitals like that feed upon the
work around them. There was work around that hospital;
there were jobs, and decent houses, in that midsized Colorado town where my father had chosen to live, because
there are still places like this in America, where the bills that
are sent are paid in full, and the struggles are more secretive.

I walked in through the ER entrance, and it exhausted me instantly, because it was huge and clean, and well equipped, with only a couple of doctors on duty, doctors who were paid well and ran on the treadmill for the money. It was the sort of place that loved the numbers, the satisfaction scores and the speed, the kind of place that advertised the brevity of the wait, and put up gleaming billboards welcoming new specialists into town.

A nurse gave me directions to the information desk, and I walked out of the ER, past the atrium, where a high school girl was playing the violin, to where the elderly volunteer sat at the desk. She beamed at me with her watery blue eyes when I asked the way, and then I found the elevators, and the hall, and the private room, and I knocked on the door.

My father looked small, and shaky, and alone. He was sitting in his hospital gown in a chair beside the bed.

● ● ●

When I was a child living in Brazil, I once got jumped by a gang near our apartment. There were five or six of them. I was going to the supermarket to buy milk for my mother, and they came out of nowhere. One moment I was walking across the grass of a park with a canvas shopping bag I'd fill with milk, and the next I was surrounded by them, laughing and making high-pitched cries, and one of them I remember slipping off his flip-flops and doing capoeira moves, as if in anticipation of a treat.

I was able to break away, across the street, running into traffic. The cars swerved and slowed, and the drivers leaned on their horns, but I got across, and into the supermarket, with my bag, and they did not follow.

I bought the milk. I waited for more than thirty minutes, peering carefully out the door of the supermarket, and I thought they'd gone.

So I crossed the street again, with the milk, and walked back toward our apartment, and for a few minutes I thought I was safe.

But they were waiting anyway, behind one of the apartment buildings, waiting because they had nothing else to do, because they were bored, waiting because they knew I was something a little different, a foreigner and a curiosity.

So they were on me again. They kicked the bag of milk out of my hands, and it fell to the ground, and the milk spilled, and then they were all over me, and as I fought back as best I could, and swung at them, I had the precise understanding that they might hurt me very badly, that this was not a schoolyard but something altogether larger and more terrifying than anything I had ever experienced before.

My arm was burning, as if it had been branded, and I realized that I had been bitten, that one of them had buried his teeth into the flesh of my upper arm, and the blue bruises of his teeth, in two perfect half-moons, stayed there for weeks as the wound healed. I remember jerking my arm free, and trying to hit back and get away, as if from a pack of wolves, and how impossible it was, because there were too many of them.

I was saved by a Catholic priest. He was walking past, in a cassock. He was young. And he came to my rescue, with his cross lifted in his hand, shouting, in Portuguese: *Stop, in the name of God.*

Miraculously, they stopped. They listened to the priest. To this day it surprises me. They stepped back and let me go, and I remember how the priest stood there, in his triumph, with a light in his eye, as if the truth had been confirmed.

I ran home, shaken and afraid, and I had a bleeding human bite on my arm, and no milk.

My father was in his mid-forties then. And when he saw me at the door, and I told him the story, he did what I would not have done now.

He pulled me back down the stairs and asked where they were, and then he went after them.

The priest was gone. But they were still sitting on the grass under the trees in the park, and when he saw them, he didn't hesitate. He just went for them, sprinting full out across the grass.

At that age, my father was still coiled and fast and strong. He wasn't that far from his youth. He'd been a boxer and a wrestler and a college quarterback. But I'd never seen anything quite like that ferocity in him before. I knew it was there, of course. But I saw it then, very clearly, and it was frightening, because he was not in control of himself.

When they saw him coming in vengeance across the grass, they broke and scattered into the apartment blocks around us. In that moment they were boys, and reacted like boys. I'm glad for it now, glad for everyone that he didn't catch them.

• • •

So I sat there with the gawky and confident cardiologist, tall with his stoop and his slightly faded white coat and his running shoes, a man my own age, who told me he liked old guys like that, who went hiking on their own in the wilderness and wouldn't listen. I knew what he meant—those with spirit, old men who won't give up and need to be dragged unwillingly from life.

The cardiologist was patient, sitting on the spare chair, explaining the particulars, answering the questions again, and my father sat there, sitting up straight, and looking at him very carefully and alertly, this man who'd saved him.

I thanked the cardiologist outside, in the hall. I heard the gratitude in my voice, and I was on the brink of tears, though I don't think he realized it.

I felt his power as he stood there in the hall, as he described the lesion to me in technical terms, almost, but not quite, collegial. And I felt the power of the hospital around me also, and the weight of the fates within it. Of course I've understood this all along, but I was used to standing where he was standing, drifting above the events, feeling their significance brush coolly against me and flow past. Now I was on the other side; I had crossed the river, and was with the others.

● ● ●

My father insisted on driving himself home. He'd left his truck in the parking garage two days before. He could buy a new one anytime he wants. But he refuses to.

It's an old truck, with almost three hundred thousand miles on it, more than ten times the circumference of the earth, and just a little more than the distance from the earth to the moon. I remember watching one of the lunar landings from his shoulders, as a very young child—I remember him looking up at the nighttime sky, and pointing, and telling me that men were walking there.

So I followed, out of the parking garage, and up the street, and onto the highway, and out of town on the overcast day, into the country, as the heavy snow-laden fields passed by, out to the house he'd rented, the house he was reluctant to show me, in its modesty, and, in his mind, in its purposeful reduction, because he has money, he has more than enough money, and he was leading me to a choice rather than a necessity.

The house was small, at the edge of a field, with a view of the mountains and an iced-over pond, and his landlord lived across the driveway in a bigger house.

His house was rented empty, so he'd picked up a few

pieces of furniture from a thrift store, a couch and beds, a few dishes and plates, just enough to get by and no more. The place was clean; the heat was on. There was soap in the shower and the bathrooms were fine. There were no clothes on the bedroom floor, and the bed was made, and there were sheets on it, and for all of this, I was relieved.

My father is a wanderer. He tries to look outward. He's chosen distance, and the clarity of dispossession, and the shedding of attachments. I am not sure that he understands why he has done this, just as so few of us truly understand why we do what we do, lost in the maelstrom as we are.

But on the walls, and on the bedside table and the dresser, were photographs of the past. My brother and I as children and later as young men, and the many places we had lived and traveled as a family, when we were a family. There was Japan, and Brazil, and Iran, and vacations to Africa and Nepal and all the rest, the Amazon and Pakistan and India, on beaches, on ski slopes, on mountainsides, on buses and in stations, the long and unconscious journey that for him continued still. And there were our children as well—my son, and my brother's daughters.

I thought of how the ghosts of the past keep endlessly rising and claiming us again, those dark angels that love us forever.

· · ·

I stayed with him that weekend, and went to the supermarket with him as he bought his dinner at the deli, and then returned to his house, and he put on classical music exactly as he had done throughout my childhood, and we spoke about the usual things, as we had done throughout my childhood, about world events, about politics, and he talked to me about art and aesthetics, and about his hiking club, and about the birds he'd seen, and about his most recent trip, and the six jaguars on the bank in the Peruvian Amazon, and he showed me the photographs he'd taken of them from a canoe, which I'd seen a few weeks earlier when he emailed them to me.

But it shook him, I knew, his near miss, as he realized that the pain he had been ignoring, the breathlessness on the path, was gone. I wondered if the solitude he had chosen, which is to say the solitude that his nature had required him to choose, was comforting or terrifying. Because I'd noticed something else: even then, as he could feel his own unsteadiness on the stairs, and how tired he was, alone in that rental house, another way station after a lifetime of way stations, he seemed relieved. It was as if some deep unspoken pressure, or deep unsettlement, was gone. He seemed, if not happy, then at least happier, in those empty rooms, as if embracing that emptiness had finally allowed a kind of peace to settle around him, even as his restlessness remained.

I know it as well.

● ● ●

I had to leave, because I had to work the next day, because my own attachments were waiting, as they do, the responsibilities we assign to ourselves, and the choices we make, and the paths we choose to follow even if we do not see them clearly.

So I drove back home, in that $40,000 car, with the heated seats, in the evening, and it started snowing again, and I should have stopped, but I kept going because I had work, because there were bills to pay, and my name was on the schedule, because remaining in that house was difficult, because my father was an old man and I didn't know what I was going to do, or how I would help him, because I know he wants me to help him, or how I would preserve myself while helping him, or navigate any of the questions I have seen so often in the faces of so many of the families before me. The years pass. Their mother has fallen again, and wants to go home, or their father is confused, and forgets them, and heats his frozen dinners on the stove and leaves it on. A brother, a sister, and the struggle to help them, because everyone works. On it goes, the same story again, and now it was mine also.

It was the same drive back as well, in the dark, in the snow, and its illusion of convergence, of straight lines pointing to the center, as I sat forward in my seat, and slowed down, and peered ahead, and thought about the past again, of my own damage and confusions, and of my own son, and how I too hope that when my time comes, he will escort me.

VERTIGO

THERE IS A HALL BESIDE THE EMERGENCY ROOM THAT IS ONE hundred yards long. It's perfectly straight, begins in the ambulance entrance and ends at the lobby. Heavy automatic doors open off the hall in both directions. Banks of elevators descend to it. The ICUs are directly above it. The radiology suites—CT scanners, MRIs, X-ray machines—are on one side, and the trauma room and the ER are on the other.

People cross back and forth, and patients are wheeled to the elevators and the scanners, but often the hall is empty. When you stand at one end, and the doors are open, you can look down its entire shining length.

At times, late at night, I've been mesmerized by it—the sense of silence and distance and perfectly straight convergence, like a mine shaft, descending.

A great many people have been wheeled down that hall. I've thought about it, standing by the ambulance entrance, watching the gurneys go by. But the hall is at its best when empty and quiet. That's when it means the most, when it feels a little bit sacred. We walk in, the deep envelops us, and the openings pass one by one.

•   •   •

Plaques hang on the eastern wall. There are dozens of them, and they're all the same—awards from management. Distinguished Nurse, Distinguished Tech, Distinguished Clerk. They are participation trophies, acknowledgments of service.

Each plaque has a photograph, and a year, and a gold seal of achievement. Doctors do not do this for one another. Doctors come and go in their facelessness, and are forgotten anonymously. But the nurses are forgotten by name. The hall is long enough, and the walls are empty, so no one takes them down. They are like tombstones in that way.

At night, in the slow moments, I've looked at the photographs many times. They stretch back more than twenty years, and I know all the faces, and now I'm one of the few who does.

A large hospital over time is full to the brim with ghosts. So many come and go, so many students, so many residents, so many patients, so many nurses, so many births and deaths—the transience flows breathtakingly around you. Only the old building is the same, the building you remember. Only the cafeteria is the same.

I've worked here for almost twenty-five years. This

rough desert city, out here in the periphery, has always felt far away, and maybe that's why I've stayed, looking at the same three ancient volcanoes out in the desert at the edge of town, primeval and stark and reptilian in their indifference to time. The stars at night are wild out there, sound carries across the empty desert, you can see the lights of the city. But to be a stranger in your own life has a certain beauty to it also, a certain astringency, as you walk back and forth and back again, across the same street to the parking lot, because the street at least is unchanged.

• • •

There are Ute and Jerry and Jamie, who are dead, and there is Charles, who is retired now, quiet as ever, and there is Mary, who is sick but working bravely on, and Erin, with her daughter, and Kate, so beautiful, who moved away somewhere, and there is Candace, and Lisa, who moved happy to California, and there is Terri, who wore her daughter's photograph printed on a button on her lapel, and Mike, and my friend Victoria, and I knew them all. Not well, not closely, not always their last names even, but I knew them as most know us, which is to say I'm passing by, and passing through, and let me lift my hand and wave.

I remember Ute in the hall. I remember her being wheeled to her appointment, and I remember how she stood, and how she hugged me, and how she was crying, and how her face was drooping, because she had a brain tumor, and only a few months left, and both of us knew it, and made no pretense of anything different.

Ute was German. But hers was not the Germany of thin glasses and fluent English and excellences, not the Germany of discipline and refinement and progressive apologies for the sins of the past. Ute's Germany was the German working class, the Germany of cheap bus vacations, of sausages and beer and frightening collective songs, and when she was a girl, she had met an American soldier, a black man, and had come back with him to the United States. They had split years ago, and she had a son who was a black man also, young, American. Her husband, she said, turned out to be an asshole.

But Ute was blond and fair and a little heavy, so blond she looked translucent, and she was a kindhearted, gentle person, and she died in her fifties, and it wasn't close to fair, or close to all right, and she put her arms around me and cried, because we had known each other for a while, for years, and I'd taken care of her son a few times.

And Jerry, long enough ago that no one remembers, that no one talks about him, of his time in the navy, or his crisply combed hair and the smell of cigarette smoke, so wiry and thin, with his little wheeze of COPD and his humor and his sly competence, because he was an excellent nurse, he had judgment and experience and a hint of darkness about him that one night overcame him, and that's all. One night, one moment, he was gone.

Then Jamie, who somehow discovered that he could wear a Scottish kilt to work without violating the dress code, and so would come in his kilt, and his tinted contact lenses, and his harmless cheerful strangeness, and his brown ponytail, such a heartbreakingly young guy, who one night out past the volcanoes, in the dark and in a storm, on a flight that other crews had refused, went into the ground in a whiteout, and the helicopter burned a little black circle in the sage. I know this because I saw the photo in the paper.

I've thought of that flight, how unforgiving it was, out there over the empty desert in a snowstorm, with the winds, and the night-vision goggles, and they couldn't get out of the weather, and the pilot didn't, or couldn't, get them down. How dizzying and how terrifying it must have been, as the downdrafts came in the dark, and the wipers went back and forth in the void, where no one would ever learn exactly what happened.

I've flown in a helicopter at night over dark ground. You can't see anything.

• • •

But most of those on the wall are fine. Most are living out their lives like all of us live out our lives, bit by bit and day by day, taking the pleasures as they come. They're raising their kids, doing their work. I know this. I need to remind myself of it. I need to see that also. They pass, plaque after plaque, beside my right shoulder.

I'm walking to the cafeteria at four in the morning, because that's when it opens. The cafeteria is in the main hospital, separated by a walkway from the ER. It takes about five minutes to get there.

I walk almost the full length of the hall. I enter it from

the trauma room, which is near the ambulance entrance, and leave it in the ER lobby, where people wait. There are always people waiting. It is empty only at the rarest moments, like snow on Christmas morning.

They sit in their plastic chairs, hunched over. There are only a handful. A homeless man, reeking in a wheel-chair, sits parked near the door.

I walk past them all, and out of the ER toward the main hospital. I'm carrying my phone and the trauma pager, and I hope neither one goes off, because sometimes it is pleasant to walk through empty halls at four in the morning, when no one is there and everything is quiet.

I cross the sky bridge with glass on both sides. At night, you see your own reflection in it as you walk.

For a few moments, I'm walking with my own ghostly imperfect twin, passing through the black glass. The glass is cold because it's winter.

The lights overhead are fluorescent. There is the faint-est hint of trembling within them, like a sound too high to hear.

Then the walls are white again, twisting and turning, and then I'm in the cafeteria.

* * *

A little crowd has gathered because everyone knows what time the cafeteria opens. There is a small line at the grill.

The cook is Mexican. She speaks English with a thick accent.

"OK, Doctor, what would you like?" she asks. She's

about my age. I don't know her story at all. I only know that she, too, has been here for a long time. The handful of residents and nurses around us wait patiently for breakfast.

"The tortilla sandwich," I say. "With green chili."

She nods, and spins the tortilla out onto the grill, and cracks two eggs beside it, and places a spoonful of cool chopped green chili beside them. I wait. She takes other orders. She has been doing this for decades, and it is effortless for her.

I watch as my eggs begin to crackle, and the tortilla begins to brown, and the spoonful of green chili begins to lose its shape in the heat.

● ● ●

I return the way I came. I carry the white paper bag with my tortilla sandwich like a child going to school.

White walls, twists and turns, the stairs, the bridge again with my reflection, and now I can see the bag, too, flashing in the corner of my eye, in my right hand, as I walk.

A corner, and another, and an automatic door, and another, each one blinking and grunting open as I wave my badge at the detector.

Then I'm in the ER lobby. The man in the wheelchair hasn't moved. The others are slumped and still. Nothing is happening. It looks like a bus station. It looks like a place where people wait for no good reason.

They are there to get out of the cold. If it's quiet, we let them. There are flakes of snow falling through the lights at the entrance. But the lobby is warm, and they're dozing and enduring, as usual.

I wave my badge a final time, and leave the lobby, and step into the hall again.

All of the doors are open. I stop for a moment, as I always do. It's empty and beautiful, and it hints at another world.

So the challenge is clear to me. The challenge is to look into those depths without despair.

II

Let the world burn through you.
Throw the prism light, white hot, on paper.

RAY BRADBURY

THE BICYCLE

We lived in London. I was six years old. I wanted to ride a bicycle.

Looking back, I realize that my parents didn't have enough money to run out and buy me a bicycle. My father was a graduate student. It was too extravagant; I would outgrow it. There was nowhere to ride it anyway, except for the park. And we were only in London for a year. I would receive a bicycle when we returned home.

Young children don't understand these things.

There was a boy who lived nearby. I can't remember where he and his mother were from. But they were foreigners also. Italian, possibly, or French—it's a detail that's lost. But they spoke English with accents that were neither British nor American, of that I'm sure.

The boy's name was Eddie. He was about my age. He had a bicycle, and had learned to ride it.

The bicycle was red. Its wheels were solid black and spokeless. I remember it with impossible desire.

So my mother spoke to Eddie's mother, and she lent us the bicycle sometimes. Eddie, in his experience, had lost interest, because when you can ride a bicycle, it is not that much fun to pedal around an English park alone in winter while your mother watches.

I remember the park. I remember leaves on the ground and rain in the trees. I remember my excitement, wheeling the bicycle from Eddie's apartment to the park with my mother, and then getting on while she held it steady.

As I recall, the bicycle had a single training wheel. Not two, only one. And so I learned to ride it leaning over to one side, like the tilt of the head. My mother would walk alongside me as I pedaled.

It doesn't take long for a six-year-old to learn to ride a bicycle. It takes only a day or two.

● ● ●

Eddie's mother, I think, was a beautiful woman. She must have been in her early thirties. She had two children— Eddie, my compatriot, and an older boy, who is hardly more than a shadow. I believe he was thirteen or fourteen, and I forget his name.

I think, but I am not sure, that Eddie and the older boy were half brothers. I know that Eddie's mother was unmarried at a time when this was uncommon. I remember a faint sense of stigma, but I don't know why.

Sometimes Eddie and I would play together in the park, and sometimes we would play with his toys in his apartment. I don't remember his toys, only that they were plentiful.

Eddie was not rich. That, somehow, I knew also. Despite the toys, and the bicycle, they were struggling. I think now that the toys, and the bicycle, had come from elsewhere, and had been gifts from others—Eddie's father, perhaps, who was not there.

Eddie's mother had come to London with her boys to be discovered. She was an opera singer.

* * *

My mother and Eddie's mother were nearly the same age. They each had young children. I don't believe Eddie's mother had a job, though I'm not certain. My mother was not working then.

So they became friends. I don't believe there was ever a great bond between them. I am not sure how they met. I know only that they were neighbors, that sometimes we would visit her, and that the visits were brief. Neither of them had been in London long, and neither of them had met many people.

But Eddie's mother had a plan to be discovered. She had rented a hall and hired a pianist. The hall was in downtown London, in the West End, and I believe it is still there, and that it is well-known. It was undoubtedly expensive. She sent out mailings, and she posted flyers. She called newspapers, and she called music critics.

She invited my mother.

• • •

I remember my mother dressing me up. Mothers like the chance to dress up their children, and this was an opportunity. An opera recital in the West End of London—it must have felt exciting, to go watch a friend take the stage, under the lights. She was going to perform a selection of famous arias. She was going to show her range and her voice and her beauty to the city.

It was over an hour of trains and buses to get there. It was dark also, because the performance was scheduled for the early evening.

My mother and I arrived early. Eddie was not there. But the older boy, dressed in a tuxedo, ushered us to our seats.

The hall was empty. There were hundreds of empty seats. We sat close to the stage, and I remember, as the minutes passed, how my mother grew increasingly nervous, as she looked around her. I have a distinct memory of her whispering to me: *I hope more people will come.*

The wait was excruciating. But no one else came, no one at all. It was just the hall, and the older boy, and his mother somewhere backstage, giving it a little more time.

Finally, the pianist she had hired walked out in his tuxedo to the grand piano on the stage. He was an older man, with white hair.

By then I understood: I was not to move. I was to sit in my chair and pay attention.

He began to play. And then Eddie's mother stepped

out onto the stage in a ball gown. A spotlight came on, and the house lights dimmed.

She performed the entire program. She sang aria after aria into the empty hall. It was more than an hour, my mother remembers. I remember it, simply, as being a long time. During the intermission, she remained backstage.

My mother was stricken. Rarely have I seen her so upset. And even then, at that young age, I understood humiliation as well. I was not oblivious.

She was a good singer. She was not insane. She'd trained at a conservatory. She did not sound out of place on a stage.

●   ●   ●

That moment has come back to me many times over the years. She could have walked out on the stage to the sea of empty seats, and realized, *There's no point. I shouldn't put myself through this.*

But she'd paid the money, money that I don't think she had. She was in her rented gown. Her son was the usher. My mother and I were the crowd.

So she did it; she sang the program for us, because we alone had come. I'm sure she sang for herself as well, out of the strength that we all need and must find. I've thought about that courage also, as she stood there in the spotlight by herself and tried her best.

Sometime later—days or weeks, I cannot recall—I rode Eddie's bicycle without the training wheel down the narrow asphalt path in the park. I remember my exhilaration. I remember the smell of the air, and the wet leaves, and the cold, and the watery English winter twilight, and my mother's presence beside me, and the feeling of the bicycle moving easily beneath me as I balanced so well.

THE REDEEMERS

His arm was red-hot from the infected dog bite, swollen to twice the size of the other, and he had a fever, and he said it wasn't the dog's fault, because it was dark when he'd sat down on the couch where the dog had been sleeping, and he was just doing his job, his dog, because he was protective. I asked him if the dog was a pit and he said yes, but the reputation is unfair and he was a good dog and had felt bad about biting him, he could tell.

So we put him in the observation unit for IV antibiotics, and by morning the swelling was down and the redness was nearly gone and we sent him home.

He showed me a picture of his dog on his phone.

• • •

She was vomiting and shaking, and she had a fever, and her lower back ached, and tapping it hurt her. I remember that she was in her thirties, and her husband and her son were with her, with identical looks of fear on their faces as she retched into a bucket, but after she'd stopped vomiting, she sent them home and told them she'd be OK, they needed to get some sleep because of school and work.

I waited for the urine to come back, and then it did, full of white cells and blood, but I'd already started the antibiotics by then.

I told her she had a kidney infection, and she nodded, because that was what she thought, too.

In the morning, she was cool and steady and the pain was gone and she could eat.

"I love my boys," she said, "but they get so worked up."

•   •   •

He was a high school kid, a soccer player, and his ankle was dislocated, and his foot pointed sideways on the gurney. He was pale and horrified, doing his best to be brave, waiting for his parents to arrive. They gave us consent, because he was a minor, although he gave us consent also, and so we pushed the drugs and he went out for a few minutes, eyes half open to the ceiling as he breathed. Then the orthopedic resident took his foot in his hands and twisted, and we all heard the sound as it snapped back into place.

"There's no fracture," the resident told the parents. "He's lucky."

A few minutes later he woke up, and scratched his nose, and muttered something incoherent about his brother, and a test, and everyone laughed, and then we put him in an orthopedic boot and sent him home with crutches.

• • •

She was clutching her belly in misery, a housekeeper in one of the hotels, and she didn't speak English, and was probably in the country illegally, and it looked like her gallbladder, and sure enough it was a gallstone, there on the ultrasound, casting its shadow, and so we gave her morphine until the pain eased.

Usually they go home when it's a gallstone, but it was a big stone, stuck in the neck, and it wasn't going to pass, so I called the surgeon.

"OK," he said. "We'll keep her and do it tomorrow."

"Gracias, Doctor," she said, her pain gone for the moment, as they wheeled her upstairs.

• • •

He sat panting like a hummingbird, and hardly wheezing at all, and he couldn't talk, and that's frightening in asthma, and as we rolled him out of the lobby to the resuscitation room, he began to get a little wild and a little desperate.

No one wants to intubate someone for an asthma attack. You have to get the tube in right away. Their lungs are so tight, and it's so hard to bag them, and their oxygen levels plummet when the drugs flow, and usually they're young, and the stakes are always enormous, because no one should ever die of an asthma attack.

But we got the line, and put him on the face mask, and gave him the drugs to breathe—albuterol, and Atrovent, and oxygen. We gave him the steroids, and the magnesium, and finally I ordered epinephrine, because he was young, and his heart could take it. The nurse gave the injection beneath the skin on his shaking upper arm, pale and covered in goose bumps.

The epinephrine turned him toward us, but then the other drugs began to work as well.

Suddenly he was wheezing everywhere, and it meant that he was getting better, because asthma at its worst is nearly silent.

"Sorry," he said, able to speak at last. "I ran out of my inhaler."

● ● ●

He was an old man with a fever and a cough, sweaty and miserable under the lights, and his wife sat beside him, and he coughed again.

Most coughs are superficial, most coughs are the tickle of a feather, but his was deep and primal and flung green mucus into the tissue paper, threaded with little strings of blood. He opened the tissue paper and showed it to me, as if he was revealing a secret.

He had pneumonia, and his right lung was full of it, and pneumonia is nothing to trifle with in old men.

So I started the antibiotics and admitted him, and he waited for hours until finally a bed opened, and I didn't think about him again until his chart came back to me.

They'd kept him for four days. The antibiotics did their job. He never got worse, and then he got better, and then he went home.

• • •

She was an old woman, and she was a little forgetful, and she lived alone.

A rash covered her body, a fine red tracery on her chest and belly and face, and it frightened her because she'd never had anything like it before.

Rashes can be anything, but it looked like a drug rash, it looked like one exactly, and so I asked her if she was taking any new medication, like antibiotics.

"I'm taking that medicine for my foot," she said.

So I looked her up on the computer, and she was on Keflex for an infected toenail that a podiatrist had removed, and she'd forgotten to tell us because it hadn't occurred to her.

I examined her toe, and it looked fine; it looked like it was healing well.

"It's the Keflex," I said. "You're allergic to it. You need to stop taking it. The rash will go away."

She looked at me in astonishment.

"Oh," she said. "I never thought of that."

• • •

He was on a roof, and came off the ladder, and went down to the concrete, and he was gasping and moaning and clutching his leg because his femur was broken, and his leg bent in a rubbery way as we slid him from the gurney to the bed in the trauma room. He was in his forties and looked like work and family and money wired home to Mexico.

By the time the orthopedic surgeons arrived, the CT scans were normal.

They used an electric drill with a long steel pin. You numb up the skin and drill the pin right through the femur and out the other side, just above the knee. I've done it a few times because it's not difficult.

The pain is bearable, but it looks ghastly. The drill is almost exactly like the ones used in the construction of houses.

We medicated him, and the drill whined; the pin went through, rising grotesquely through the skin on the far side before breaking the surface, and then they attached the pulleys and pulled his femur straight, and loaded on the weights, and after a while he was comfortable again, lying there looking at the pulleys, and pin, and the weights, and the little strings of blood running from the pins.

"OK," the orthopedic resident said, finally, grudgingly. "We'll admit him and put in a rod tomorrow."

So I stood next to him and explained, in Spanish, what was happening. I explained the surgery, and how he would walk normally again, and that it was a common thing. In a month or two he would be back at work. I asked him if he wanted more morphine for the pain.

He shook his head.

"No, Doctor, thank you," he said, in English. "I am OK."

●　●　●

He'd been slashed in the belly, and a little gray loop of small bowel lay there in the light, and its presence made everything easy. Everyone knew what needed to be done.

His arms were crossed, and he stared across the room in fury, and he wasn't talking. He had a teardrop on his cheek and a large green "505" tattooed on his abdomen. The loop of small bowel lay there on the numbers.

"Did you do this yourself?" I asked, but he snorted.

"No fucking way," he said, so I believed him.

"How did it happen?"

"It don't matter," he said, and was silent.

We all stood there for a few seconds, looking at the glistening bowel

Then the trauma attending walked in.

"Do you want a CT scan?" I asked.

He thought for a second.

"Nah," he said. "We'll just take him up."

So they did; they took him to the operating room, and opened his belly, and washed everything out, and his bowel was somehow uninjured.

He went home also, a few days later, still refusing to tell the story, following the code like a soldier.

●　●　●

He was in agony, and he paced and moved, and bent over for a while, then straightened again, and paced some more, and dry-heaved a little, and it looked like a kidney stone from across the room.

We gave him Toradol, and then we did the scan and saw it, white as the moon on the gray screen, a perfect little circle of calcium, but it was small, small enough to pass, small enough to do nothing.

The Toradol came like a cool breeze, and five minutes later he was himself again, and his pain was a memory alone.

"Ever had a kidney stone before?" I asked.

"Never," he said. "And I don't plan to again."

We laughed, and then I got him the paperwork, and his referral, and he walked out the way he came.

•   •   •

His shoulder had dislocated too many times. But he was afraid of surgery, and so kept putting it off, and now it was coming out in his sleep when he rolled over. I walked in. I was working alone. I'd seen him before.

"So it's out again?" I said, and he nodded, holding his shoulder.

"I know I need to get the surgery."

"You're right," I replied. "You do."

When all the ligaments are stretched like that, sometimes a shoulder pops in with nothing at all.

So I sat him up straight. I took his wrist in my right

hand. I took his elbow in my left hand. I bent his elbow. I knelt beside him, on the floor, and pulled slowly.

You feel it in your hands. It's like dropping a small bag of sand on wet grass. The ball snaps back into the socket. It is immensely satisfying. They know instantly. It is a nearly perfect act.

"That was awesome," he said, because he was young.

I put him in a sling.

"Don't take off the sling," I said. "Go see the surgeon. It's going to keep doing this. The reason that worked is because everything is so loose."

"I know," he said, very seriously and earnestly. "I know you're right."

• • •

He was seizing, his back arched, his arms contorted, his eyes half open and beating sideways, and then he started shaking, and he wasn't breathing, but seizures almost always stop. So you watch, and you wait.

"Does he have a line?" I asked when I walked over. He looked familiar, but I couldn't place him at first.

"He just got here," the nurse replied as she raised the head of the gurney. "He seized at work earlier."

We watched as the seizure continued, and followed its pattern. Everything tenses, all of the muscles go tight, and then, as the seconds pass, the shaking begins, a fine tremor at first that then grows wilder before your eyes. A seizure is a crescendo, like a piece of music.

But then it slows, and finally it stops, and they go slack, as if swept clean.

The nurse pushed the Ativan.

It is always hard to judge the time because time passes slowly when you are watching that. But a few moments after the drug went in, the seizure stopped.

He breathed, deeply and coarsely, for a while. A few minutes later he began to wake up, moaning, as if in sleep.

I went back to the desk and pulled up his records. It was then that I remembered him.

A few minutes later, he was sitting upright in the gurney, back in the world and alert.

"You've got to take your Dilantin," I said. "How many doses have you missed?"

"I'm supposed to pick it up tomorrow."

"Come on," I said. "You've got to take it. You know that."

He sighed, and after a while he nodded.

"I know it's fucked up," he said.

He'd had epilepsy since he was a small child. He was young, and wanted a young man's life. He resented his epilepsy, and resented his Dilantin, and I understood this.

I thought of the boy in the river, all those years ago.

THE DWARF

IT WAS RIGHT ABOUT CLOSING TIME, NEAR MIDNIGHT, WHEN they brought her in, and I heard it on the radio before the trauma pagers went off, and I could tell from their voices, and I felt it again.

That feeling is best suited to the hearts of the young. You know that something frightening is coming, that it will be all over you in a moment, that there will be no time to think. You never get used to it. You only become familiar with it.

When they're dead, when they are damaged beyond repair, when their pupils are wide-open and their bodies are loose and empty, you can see it at a glance. Calm descends upon you. You know that the story is already at an end, and that everyone standing in their gowns and gloves is an actor, performing last rites. There is the comfort of distance, and the illusion of mastery. When you know you're powerless, you feel a certain power. When they are out of reach, they can't touch you.

But when they are still alive, when every act matters, when you know that you will be neck-deep in the icy torrent of events rather than watching them flow past—that is when the feeling comes down upon you. When you've heard so many radio calls, you can tell.

The paramedics were afraid. Their voices were rising.

● ● ●

I'd never seen anything like the woman on the stretcher before. She was the height of a child, but the weight of a grown woman, and she was gurgling and unconscious and blue, with lipstick and dyed-blond hair, and I could smell both the alcohol and the vomit in her mouth. Her thick white arms were speckled with needle holes from the paramedics' failed attempts.

"We don't have a line," they said, rushing. "Her sat is in the 60s, she's aspirating, she's hard to bag."

"What's wrong with her?" one of the nurses asked as we moved her from the stretcher to the gurney, and turned on the suction, and sat her up a little, as her head rolled back and forth.

"She's a dwarf," the paramedic said. "She was dancing on the bar and fell off."

● ● ●

We had to intubate her, and her neck was thick and short, and her face was misshapen, and we didn't have an IV, and the ridiculous was not the ridiculous but the terrifying. Every outward sign was pointing at darkness. Obese, with a mouth full of vomit, and a fat tongue and lips, and no landmarks in her neck for a scalpel, hypoxic already, when every squeeze of the bag pushed more vomit into her lungs, as she struggled incoherently against us, head-injured and drunk, a strange wild creature who had somehow placed her life in our hands.

Situations like this are like hurdles, and you jump, and you jump again, but sooner or later you know that one will knock you down, that you will fail, that for days you will have shame and sleepless nights. That passes also, but you never forget, and each case like it is another rung in the descent.

The nurses got a line. How, I'm not sure. But they did, a tenuous one but flowing, in her upper arm, white as the belly of a snake. We were suctioning her and bagging her, but she was biting on the Yankauer, and we couldn't get it between her teeth, and vomit was bubbling from her nose and splattering us, and we had to jump.

So again that same exhausting decision, to push the drugs, the paralytics, that stop everything but your heart, where you can't breathe at all, where you die without the tube, and it was foreign ground, her throat, and the light on the laryngoscope was a tiny point in the dark.

The drugs went in. She stopped breathing. The numbers on the screen fell like a stone.

"Try to bag her," I said to the resident, and he did, pressing hard and lifting, and the numbers slowed in their descent for a little while, but they didn't rise, and he stood there with the respiratory therapist, adjusting and lifting as the seconds passed.

"Try the LMA," I said, and he slid it in, a fat plastic tube with a diaphragm on the end, the shape of a leaf, that you press in close to the trachea, and attach to the bag, and hope will make a seal. Usually it works.

But it didn't work, and wouldn't fit, and the seal was juicy and wrong, and each squeeze blew little bubbles of vomit out of her mouth, and I knew my voice was rising.

"Take it out and bag her."

So he did, pulled it out like a wet green root and threw it on the floor.

"Just try," I said, and he did. He took the video laryngoscope and put it in the woman's mouth, and I looked on screen, and all I could see was a greenish sheet of fluid among the pink tissue.

"More suction," the resident said, and we got another Yankauer, and they hissed, and he struggled with the blade, and I had my hand on her throat, and we couldn't see anything.

"Sat's falling," the respiratory therapist said.

"OK, stop, bag her again."

So the resident took out the blade, and pressed the mask to her jaw again, and we squeezed the bag as hard as we could, because we were rough and frightened then.

"Sat's coming up," the RT said, and it was, rising from the 60s to the 70s and then, barely, to the 80s, where it remained.

There is a threshold you can't cross. After a while you can feel where it is in patients. You know you are close; you know the heart is about to stop. We were at the threshold.

"Call anesthesia," I said, which is an admission of failure in emergency medicine, and I knew I should have said it from the start, but I was behind the events rather than in front of them, and reacting, and she'd been there hardly any time at all.

The nurse ran to the phone.

"Let me try," I said, because I had to, because it was my job, and my responsibility, and all the rest, and I thought I'd fail, because I'd seen the screen, and that anatomy, and the vomit, and my heart was pounding, and the resident was good, and knew this procedure as well as I, and he was stronger and younger, and could lift the jaw with the blade more forcefully, and I was rusty, and all of this I knew as we exchanged positions.

"Anesthesia is on their way," the nurse said.

But "on their way" was at least five minutes, and likely more, even if they ran for the elevators, and they never run for the elevators.

So I tried. I put the video laryngoscope in her mouth again, and then the screen went gray and misty. I pulled it out, and wiped the lens on the bedsheet beside her, and tried again, with the suction, and then the lens clouded over once more, and I took it out, and wiped it on the bedsheet, and did it again as the numbers fell once more, and anesthesia was nowhere to be seen.

On the sixth attempt—later, I counted them—I got

a glimpse, like the parting of fog on a mountainside, just enough to reveal the path. I slid the thinnest tube I could toward the place I thought her vocal cords might be, and somehow it went in.

"I'm not sure," I said as the respiratory therapist attached the bag to the tube in my hands and began the slow squeeze.

"I think you're in," he said, and I realized that I was. I realized that the numbers were rising, and that we were back.

It is easy to claim such moments for oneself. But I know better. Events respect neither strength nor weakness, no more than they respect intelligence, or ignorance, or wisdom.

When anesthesia finally showed up so casually with their little orange box of devices, she was in the scanner and alive. I said nothing, but I was angry, because they'd taken their time, in their arrogance—as if to say, *We are above this, and urgency to you is not urgency to us.*

● ● ●

The jokes began in the CT scanner as we waited for the images to come up on the screen. Dwarfs, bars, dancing, and the rest, and it was funny in the moment, in our elemental relief, and suddenly we all were laughing back there, behind the window, watching her on the table.

The table inched forward. The numbers on the monitor were blue. The images came up one by one as that beautiful and powerful machine did its ghostly, silent work.

Her brain was clean and gray and full. A young woman's brain, undamaged and perfect, as if suspended in space.

The next morning, she woke up. They took out the tube and sent her home. She remembered nothing.

So occasionally I think about her, the woman whose life I saved. I remember her lipstick smeared on my gloved hands, and the vomit, and the vodka. I imagine her dancing in a bar, or at a party, and I imagine her misery, and how difficult it must be for anyone, to have that affliction. But I imagine her joy as well, when the crowd is smiling with her rather than against her, and cheering her on, as sometimes, surely, they do also.

THE MACHINE

THEY WERE COMPRESSING HIS WIZENED GRAY CHEST ALL THE way down the hall, and he looked dead from the start. Thin, already intubated, with tufts of a scraggly white beard, smelling of cigarettes, and his head rolled loosely and dangled for an instant when we moved him over from their gurney to ours.

They'd found him unresponsive on the ground. No one knew how long he'd been out. No one knew anything about him. He was just an old man, a smoker by the look of it, laid out by time like the rest.

When they began, his heart had not been beating. But in the ambulance, they'd gotten a sign. His heart had started again, for just a few moments, just enough time for an EKG.

They handed me that EKG. It showed a heart attack.

There is a new device. It's a battery-operated pneumatic piston that performs CPR automatically. You put a base plate under them, line up the piston in the right position, and turn it on. You can set the rate, and the depth, of the compressions.

We pushed the button.

The machine looks ghastly. The piston pumps, up and down, with terrible force. It is far more powerful than a man. The compressions are deep and hard and brutal. The chest collapses, springs back, collapses again.

Suddenly, this machine is everywhere.

"What's the rhythm?" I asked the resident, after a bit, because he was beside the monitor.

"Still asystole."

"Maybe we should stop," I said. I remember saying this very clearly.

I'm not sure why I kept going. But I did. We let the machine continue for a while longer. We kept giving the drugs. Perhaps it was the effortlessness of the machine, and the calm I felt. No one was getting tired.

"He's got a rhythm," the nurse said suddenly, and it was true. Out of nowhere, normal complexes had sprung up on the monitor.

"Pulse check," the resident said.

The nurse touched the button. The machine went silent. We felt his neck, and his groin. He had a pulse.

I had no choice.

"Activate the cath lab," I said.

So the call went out, and a few seconds later, a dozen or so pagers were vibrating on the hips of people all over the hospital. It's an expensive call. It costs $10,000. Wherever they were, the cardiologists, the nurses, the technicians— all of them paused, and pushed the button, and read the text on the screen at the same time.

Then his heart stopped again.

"Start CPR," I said, and the machine began again, like a dark thing.

The cardiologists arrived. I apologized to them.

They looked at the EKG and sighed. His ribs were broken. I could see them moving beneath the piston.

But then the complexes winked at us once more.

"Does he have a pulse?" the cardiologist asked. So we stopped again, and there it was, leaping greasily from his groin against my finger.

Codes like this seem endless. You get a whisper, a flicker. So you keep going, even as you know it's pointless. Sometimes this continues for hours. They keep coming back, a little. You keep lashing them with the whip. They go out; then they come back. You can't stop, you have to be sure, but all the while you know. You feel the weight of experience, of so many dozens of others, over the years, and you know how they ended. You still keep going. You get out the ultrasound machine and look at the heart.

"He's got activity," the cardiology fellow said, and shrugged.

For the moment, he was lying there alive. The machine was off. The ventilator hissed. All of the drugs were flowing. I thought of him bleeding from the work of the machine.

"OK," the cardiology attending said, resigned, because she had no choice, either. "Let's go."

And so they went. They rolled him upstairs to the cath lab. In the world we knew, he was dead.

* * *

Later in the afternoon the resident came to me. He was excited, with a light in his eyes.

"He's following commands," he said.

And it was true. The cardiologists had opened the blockage. I read the note in disbelief. I saw something else as well. He was only sixty-three. I'd thought he was much older.

Following commands means the brain is alive.

Being surprised means you were wrong. Being surprised is frightening. You have to learn from that surprise. You have to recognize it for what it is. It is not the grace of God, not the miraculous cure. It is your own failing to see the world correctly. You want to see the world correctly. You want to believe that your convictions are earned, that order is within you.

But I still thought he was going to die. I thought his kidneys would fail, or his liver. I thought he would bleed from the machine. His body was a bruise, a beaten thing. The young, sometimes, can take it. But the old are like flowers in the brightest sun. They can stay there hardly at all.

But I felt it, too. I felt the excitement. I felt the hope and the power.

• • •

So it became a kind of secret between us. The resident checked on him every day, on the computer.

"How's he doing?" I'd ask as the time passed.

His kidneys did shut down. They put him on dialysis. And then his kidneys came back. His heart was damaged. But it beat on anyway. His liver function tests did rise. But then they came back down. He did bleed into his belly from the force of the piston. But the bleeding stopped. He got pneumonia also. But they gave him antibiotics and kept him on the ventilator, sedated and gone from the world.

"Don't get too hopeful," I said as he kept checking.

By then I knew it was the machine. I knew it was the power of the compressions. The force that had broken his ribs and lacerated his liver had also reached like a fist down to his heart, and squeezed it like a true friend. His blood had kept flowing, just long enough.

I had not believed in the machine. I'd thought it was another of the false and expensive advances, another of the wands we wave at the end. I had stood there, in my experience, and did not understand that I was looking at a new world. Not a revolution, not a great discovery. But a step forward all the same.

With weariness comes skepticism. With every lecture comes questions. The slow crawl forward is full of lies and money. Therapies sweep across the journals, then fall away. Sometimes they are never seen again; sometimes they come back in distorted forms. Careers rise, and fall, and ambitions with them. Treatments that do not work are offered to the thirsty and the desperate in return for great sums. The patterns we know are the patterns we see. I had almost let him die.

On the twenty-first day, he rose and walked again.

THE COLLECTED WORKS OF WINSTON BEGAY

She handed me the drawing at the doctors' station.

"Winston is back," she said. "Here's another one for your collection." She winked, in her way.

"Is he still here?" I asked.

"Yup," she said. "I sent him to the lobby."

I looked at the drawing. It was like the others, in blue ballpoint pen, about the size of a postcard.

A teepee, and a landscape, with a round circle for the sun, with lines for rays of light, and a campfire, and his signature eagle feather. In the bottom right corner, he'd signed his name.

It's the nurses who do it. They give him the pen and the paper. In return, he leaves a drawing, like a little white leaf in the chair.

People like Winston have distilled solitude to its purest form. But he is not alone. The streets are full of travelers.

●   ●   ●

Mark Garcia was on dialysis. He had a chair three times a week and nothing else. It was his most valuable possession, and a precious thing. His dialysis chair cost the state tens of thousands of dollars a year.

Once, I saved him. I saved him so completely that I felt possessive of him. I'm angry at him still, because he betrayed me a year later, when he finally waited too long, and they found him dead.

But that day, when they handed me an EKG, I saw his name. I saw the pattern. I knew what it was, and who he was, and why he was there, and so I went running out to get him, in the lobby, slumped in his plastic chair.

On the way to the resuscitation room, on the stretcher, his heart stopped and his eyes rolled back in his head.

We used his dialysis port, because there wasn't time for an IV, and the nurses pushed the amps of bicarb and calcium and insulin one after the other, syringe after syringe, into the port, trying to drive the potassium down and buy time.

Finally a few normal complexes began blinking out of the flat green line on the monitor above his head, and then they all came back and he woke up.

When your kidneys fail, you can't excrete potassium.

When the level of potassium rises too high, the heart stops beating. Sometimes, you can tell from an EKG.

Dialysis is a crude kidney, as ungainly as an iron lung, and a dismal sentence, but it removes the potassium, and the urea, and the fluid. If you go, it keeps you alive.

He didn't go. He drank in the parks instead.

"Mark . . ." I began, a few minutes later, when I'd made the phone calls to the nephrologists and the emergency dialysis team was coming in.

He interrupted me. He looked guilty behind the oxygen mask and the albuterol we were giving him.

"I'm sorry," he said, because he knew what was coming. "I'll be better."

I made him endure another sermon anyway. He looked like a small boy, nodding along as I lectured him and told him he was going to die if he kept doing this, that he was lucky he was alive today, do you know how close you came? Do you, Mark?

"Yes," he said. "I do."

• • •

I was driving on a cool and sunny Sunday afternoon, on a day off, when I saw the man lying in the center of the street, not far from the hospital, in a thigh-length black leather coat. I stopped just as an ambulance pulled up and the paramedics jogged over to him, because someone must have called.

The paramedics were kneeling when I reached them, and were careful, but there was no sign of a car, no blood on the pavement or any other evidence of a crash. Just the man lying in the street as I introduced myself and asked if I could help.

They rolled him over.

"It's Doug Binder," I said, because it was. He was unharmed, breathing just fine, nearly as drunk as it is possible to be. I could never figure out exactly what ethnicity Doug was because he seemed like he had a bit of everything in him.

So the three of us—the two paramedics and me—laughed a little because all of us knew him.

Doug looked good, better than I had ever seen him. His long black ponytail was shiny and clean. His leather coat looked remarkably expensive. He had a silver necklace on, and clean black jeans, and cowboy boots, and they all looked new, and I even got a whiff of cologne. I suddenly realized that in certain conditions, and in the right context, Doug Binder was almost a sharp-looking guy. I could imagine him sitting coolly on a bar stool, ordering whiskey on the rocks.

Somehow, that time, he must have gotten his hands on some money.

They loaded him up, and slid him in the back of the rig, and drove slowly off, in the direction of the hospital, where in a few minutes other people I knew would recognize him again, and make yet another chart, and put it in the rack, where it would wait, like he would wait, until he could walk again.

• • •

Most of the time Donald Williams was harmless, mumbling to himself, but sometimes he'd take a feeble swing or two, and so the nurses watched him.

He had a way of appearing, suddenly and without warning, at the doctors' station. He'd glare, and rasp, and demand things like orange juice and lunch. He could startle you, and you'd jump, until you realized it was him.

"Go back to your room," I'd say, and point, and sooner or later he would shuffle back to his bed and sit down.

He was unique among them in that his family had not abandoned him. Each month his brother would appear and pay his bills, without fail. It was a point of pride for them, because Don was defined in his life by refusal.

He refused the help his family tried to give him. He refused the apartment, refused the facility, refused the shelter, refused the money; he refused it all. He was pure and true to himself in that way.

He refused surgery on his leg when his toes were blue, and over the next year or so we all watched them get bluer.

Finally, he refused treatment for his lung cancer, and in this at least he was correct, because there is no treatment for lung cancer.

Don was not of this world, but neither was he fully of another. He was stateless on the border. He mumbled and paced, but when he was put to the test, he could gather himself and answer the questions, and he did, with a sly look, giving the psychiatrists the room they wanted, and always want, which is to say, *Mr. Williams does not meet criteria for psychiatric hospitalization.*

So he would doze until morning, and walk out the door.

• • •

Now there's only Winston, and the new faces joining him. Younger faces, because you don't last long like that. You only have a few years on the streets, just a few, because time is different there, and passes twice as fast.

Only Winston seems the same, in his slow orbit around us, circling the city in the day and reeling back at night.

The nurses made a book and gave it to me.

"The Collected Works of Winston Begay," it read, in floral nurse script, on the white cover.

Within it were dozens of drawings, each the same size, each signed, each with an eagle feather, and a teepee, some with a peace pipe, some with a tomahawk, or an arrow, many with the sun and some with the stars. They look like the drawings of a child. But they are always about identity. They are always about being Native American.

He is Navajo. But the Navajo did not use teepees.

• • •

Melissa was excited. He was in the lobby again, and she had been waiting. She is young and blond and blue-eyed, and she's funny and wry, and she's kindhearted, and when she was nine months pregnant, she'd say—look, the brat is kicking me, feel it.

So I'd put my hand on her full belly, and feel the baby kicking, and think about life, and youth, and promise.

She made the T-shirts. She made one for herself, a few for the other nurses, one for me, and one for Winston himself.

The T-shirt is gray cotton. On it is a drawing—a teepee, a peace pipe, an eagle feather, and his name.

"He's here," she said, breathlessly, in my ear. "Come on."

So I stopped what I was doing, and left the doctors' station, and walked up to triage, into one of the little rooms, and the other recipients of the T-shirt were there, and she tried to have a little ceremony for him, leading him from the lobby, her face full of happiness.

"Winston," she said, "I have something for you. It's a shirt with your art on it."

So she gave him the shirt, and he took it, and held it in his hands, and stared at it with puzzlement; he stared at it as if he had no idea at all what he was seeing, and he was swaying, and I realized that we should have waited a little while longer, that he was so drunk he couldn't understand, or ever recognize that they cared for him, that he wasn't nothing to them, that they were not indifferent to his fate, and not unkind.

I don't wear the shirt too much, because I don't think it will hold up in the laundry for that long. I wear it sparingly. I wear it especially on the nights I think will be dangerous, the nights of the early summer, when all of the residents are new, and inexperienced, and just beginning, and everyone is a little bit afraid.

WOMEN, AT NIGHT

AMANDA, THE CHARGE NURSE, IN COWBOY BOOTS AND RED scrubs, comes up behind me at the desk. She grew up with a pack of brothers in a Western town. She puts her arms around me and nuzzles my cheek from behind. Then, she runs her fingers through my hair. I know it's her immediately because this is what she always does. It's intimate, but not sexual. The line is perfectly clear.

"Come and see this guy in 3," she says.

So I get up and follow her. She's a single mother with young kids and a bitter divorce behind her. She works nights. She's still young. Once she showed me pictures of her heavily tattooed boyfriend flexing by the pool in Las Vegas.

"That should last," I said, and she laughed.

The guy in 3 is sick. He's yellow, and his belly is distended, and his wick-thin arms are trembling.

"Thanks," I say, and she purses her lips and blows me a kiss.

"All yours," she says, and walks away.

• • •

So I send Veronica in to see him. She's the senior resident. Her family is from Taiwan. She's a planner, has a spreadsheet for her student loans, and texts rapidly with both of her thumbs.

She comes out after a minute or two, sits down at the computer, and enters a flurry of orders.

"He's in liver failure, and he's GI bleeding and maybe septic," she says. "Vitals are OK for now, but he might need the unit. I'm starting him on CIWA and giving him Vanc and Zosyn."

"Sounds good," I say. "Does he need blood?"

"The labs aren't back. I've sent a type and screen."

Everyone is quiet for a while.

Veronica calls GI. I can hear her on the phone, and know immediately that she's getting angry. She speaks in a tight, measured tone. They go back and forth awhile; then she hangs up and sighs.

"That GI fellow is such a bitch," she says. "I hate her."

"Why?" I ask.

She hesitates.

"She's Asian," she says, finally. "And she's giving Asians a bad name."

We all laugh about that for a bit.

• • •

Later, a group of nurses have gathered around the counter at the nurses' station. They are giggling and passing a pen around. Each of them, after a moment of consideration, is making a check mark on a newspaper spread out on the counter.

It is open to the Wanted page.

"What are you doing?" I say.

"We're not telling you," they say.

"Come on."

Next to the photographs of the felons in the paper, there are check marks. Some have several, and some have none.

"Come on, out with it," I say to the ringleader.

"All right," she says, and points to the newspaper. "Tell me which one you'd sleep with."

So I look at the paper. The felons stare back—armed robbery, trafficking a controlled substance, failure to appear, counterfeiting, murder, felony child abuse.

"That one," I say, and make a check mark. "I like the tattoos on his face."

They laugh, and later they start grooming one another, because nothing is going on, the pagers aren't going off, the patients around them are sleeping. They sit and braid one another's hair. They're so young.

• • •

The intern is tentative and does not speak English that well. I have to listen carefully.

She tells me about a man with abdominal pain. I listen, and know I'll start from the beginning when I go in the room. She's afraid, in a foreign country far from home. She doesn't fully understand the rules yet. I understand this. She needs encouragement and pressure alike.

"Good presentation," I say. "This is the second patient you've seen on this shift, right?"

She nods. The shift is four hours old.

"Try to see a few more patients on your next shift if you can."

"OK," she says, flushing. She rises and picks up a chart from the rack.

"Don't worry," I say, because I can see her nervousness. "You're doing fine."

She gives me a quick, tentative smile, and walks off briskly toward the cubicle.

"You have to watch her," Veronica whispers, out of the corner of her mouth. "She was my intern on MICU."

"Veronica," I say. "Do you think I was born yesterday?"

· · ·

Val comes over. Earlier, she was in an impassioned political argument with one of the paramedics about gun control. She's a physician's assistant, and has experience. She's a little older than the residents. She rattles off the presentation quickly. Bronchitis, maybe pneumonia, chest

X-ray isn't back, really I think he just wants a place to sleep. I'm going to give him a Z-Pak.

"OK," I say, and glance in the room a while later, where he's sleeping with his backpack beside him and breathing easily in the darkness.

"I can't believe him," she says, still thinking about her argument. Her passion is so reassuring, so redeeming, I think, because she's still flushed and furious even though the argument was a couple of hours ago now.

"You can't change people's minds," I say, and she nods.

"I know," she says. "But he's such an idiot."

Val is seven months pregnant with her second child.

* * *

The labs come back. The man with liver failure needs the ICU. Veronica gets on the phone. I get up, and go back to the room to check on him. He looks the same, yellow and shaky. He's awake, and I talk to him a little, to make sure he's alert enough.

He's alert enough. The antibiotics are running in.

I walk back to the desk. They're all working on their computers, typing their notes. The computers are endlessly thirsty. They generate the bills, which at times, in my dark moments, I understand as the real reason we are here.

The intern comes back and looks upset.

"What's wrong?" I ask her.

"Nothing, it is OK," she says.

But then the nurse approaches me a few minutes later.

"Your intern is a problem," she says.

"What did she do?"

"I was about to put in a Foley and told her to come back, and she got in my face and said I could do it later."

"Give her a break," I say. "Come on."

"She doesn't know what she's doing."

So I soothe the waters a little, and talk to the intern, and talk to the nurse.

"She was very mean," the intern says, finally.

I've seen this so many times. Women do not like being told what to do by other women.

Plus, nurses have an eye for weakness.

* * *

Then the pagers go off and we're in the thick of it. A man has been struck by a car.

So Veronica and I gown up and get ready, and I'm a little nervous because I know it's going to be bad. When people are struck by cars in the middle of the night, it usually means they were drunk, and stumbling into the street, and it usually means the driver didn't see them, or saw them too late, and hit them at speed.

Veronica's eyes are bright behind the mask, and she isn't saying anything. She lays out the airway equipment, the laryngoscope and the endotracheal tubes and the rest, and puts them precisely on the tray beside her. I stand back and watch. I haven't seen many bad traumas with her, because it's early in the year. I'm trying to get ready myself, mustering up the stillness that you need.

The trauma team files in. They're in their gowns and gloves also. We wait beside the gurney in silence.

Then I can hear them coming down the hall, and then they're through the doors, and I was right, it's a mess.

He's homeless, and filthy, and probably in his forties, and both of his splintered femurs are projecting through his torn jeans into the open air. He's writhing a little, and reaching incoherently, and his face and matted hair are covered in blood.

The paramedics talk—a dark street, the same story, hit so hard he came out of his shoes. They had a line, but that was it.

We drag him off the gurney and onto the bed, and everything begins. We cut off his clothes, and his twisted broken feet fill the room with their odor, in his bloody crusty socks, and then the nurses get another line. Veronica speaks in a quick, sharp voice: "Hang two units of O-positive please, we're going to intubate him."

So I watch her stand there like a razor, and say nothing.

"Do you have the drugs drawn up?" she asks, and the pharmacist answers.

"I have roc and etomidate."

"All right," Veronica says. "Give him forty of etomidate and a hundred of roc."

The nurse pushes them one by one.

"Etomidate is in," she says, and then, a moment later, "Rocuronium is in."

We wait. Veronica holds the bag on his face, fluttering oxygen around his mouth. The drugs need a little time.

"His jaw's broken," she says, because it's loose and floppy under her bloody fingers.

Then he goes limp, as they do, and it is up to us. Veronica is looking at the clock.

"Thirty seconds," she says.

"Just try," I say.

She puts the laryngoscope into the man's mouth, and I get a glimpse of blood and broken teeth, and I have my fingers on his throat, and the tube in my free hand to give her.

I can feel the tissues rising as she lifts the blade. I can see her peering down into the tunnel with great stillness and intensity.

"Tube," she says, suddenly, and I hand it to her.

She puts it in the man's mouth, and then I can feel it slide beneath my fingers.

"It's in," she says. The respiratory therapist attaches the tube to the ventilator as the X-ray tech wheels in the machine.

"What vent settings do you want?" the respiratory therapist asks.

"Whatever you want," Veronica replies. "Just leave him on 100 percent."

"OK," she says, and touches the screen as we roll him over, slide the X-ray plate under his back, and roll him back again. All of our gloves are bloody.

"X-ray," she calls, and all of us step back, away from the radiation.

"Get the pelvis, too," I say, and she nods.

Then the trauma surgery attending walks in. She's young, in her middle thirties still, and hasn't been out of training for long. She has long black hair and a steady gaze, and she doesn't talk much. I hardly know her.

We roll him to the scanner. He is still alive. His legs are crushed and twisted. His pelvis is shattered. We've put a binder on it, around his hips, pulling the bones together. The blood is running in from the bags.

The images start coming up on the screen, just as the monitor begins to ring.

"How hypotensive is he?" the trauma attending asks, sitting at the computer screen.

"He just dropped to the 70s," the nurse says, peering at the monitor through the heavy glass window of the CT scanner.

"OK, hang on a second," she says, looking quickly at the images as they unfold, one after the other.

"He's got a ruptured spleen and a liver lac," she says, then hesitates for a second. "What's his pressure?" she asks.

"Still in the 80s," the nurse answers.

She picks up the phone and dials. A voice answers.

"We're coming up," she says. "Is there a room ready?"

Then she turns to the nurses.

"We're going to the OR," she says. "Right now. Don't take him back to the trauma room."

So in a few seconds they're gone, wheeling him down the hall at the distinctive speed, with the ventilator rolling beside them, and the red bags of blood in the air like flags.

We stand there in the aftermath, in the trauma room with its signs of passage, the blood on the floor, the filthy clothes, the needle wrappers glinting in the lights.

"Veronica," I say, and put my hand on her shoulder. "That was excellent."

She blushes a little.

"Thanks," she replies as we walk back down the hall to where the others are waiting.

Later, early in the morning, we pulled up the OR note together, and saw what the trauma surgeon had done, how she'd removed the spleen and packed the liver, how she'd given unit after unit of blood and plasma, how he'd been on pressors, and how they'd taken him to interventional radiology, and threaded in catheters to cauterize the bleeding vessels in his pelvis. It had been hours of work. His injuries were terrible, and I thought it was all for nothing.

But a few weeks later, his chart came back to me. I had not signed it in the rush. I looked him up on the computer, and then I remembered him, and thought about that night again, when everyone I worked with was a woman, and how unremarkable this had seemed, and how impossible it would have been when I was young.

He was sitting up in a chair, and speaking.

LISA MADE US WAIT

"WE'RE HAVING A DISCO PARTY BECAUSE I MET THAT MAN OF my dreams and am moving to California," she said, breathlessly.

Lisa was a charge nurse. She wanted a send-off.

Lisa talks constantly. Her silliness is perfectly natural and perfectly calculated at the same time. Behind the silliness is cool assessment.

"It's not fancy," she'd say, about anything bad. A brutal trauma—not fancy. An annoying resident—not fancy.

It doesn't sound like much. But then others started saying it, too. It was an unconscious imitation, and spoke volumes.

Charisma is mysterious.

●　●　●

Most of the time, doctor parties and nurse parties do not mix. Nurses are rarely invited to doctor parties. And doctors are rarely invited to nurse parties.

Doctors don't invite nurses to doctor parties because it doesn't occur to them. As doctors get older, it occurs to them even less. This is due to arrogance. There is no other word for it. It is not a question of shared experience. That doesn't cross anyone's mind.

Nurses don't invite doctors to nurse parties, either. Nurses feel faintly oppressed in the presence of doctors, as if they can't be fully themselves. This is because doctors are often stiff and awkward and judgmental.

Lisa invited a couple of doctors to her party anyway. I was flattered.

*   *   *

I arrived a little late. Most of them were there already, at the downtown bar, and most of them had started early, and they'd been going for a while. Many wore costumes.

Melissa was dressed in a miniskirt with black fishnet stockings, and she was covered in glitter.

Christine was there also. She is half Navajo and half African-American, has perfect skin, and likes to glower.

"I'm not a Chigro," she says. "I'm not Chinese at all."

She looks it, a little.

But Andrew had gone all out. He was in a white jumpsuit with platform shoes, sunglasses, and a cape. His shaved hairless chest gleamed with baby oil mixed with gold dust, and you could see it because his jumpsuit was unzipped to his waist.

Andrew was once a homicide detective. He's held people at gunpoint and seen a lot of bleakness. He left the police due to their cruelty toward him, their jokes behind his back. He never told me this, exactly. Once, when I was feeling low, he saw it, and passed me a note of encouragement. I've never forgotten that, because you don't.

Then Jose made his entrance. He was born on the rough side, full of homies and heroin and gangs. He loves nothing more than the provocative statement.

"I'm going to ride him like Yoda," he'll say, talking about a forthcoming weekend, trying to get a reaction, looking at me expectantly.

"Shut up, Jose," I'll say, and he'll laugh. It's a game we play.

He was wearing chains and a bandanna and was covered in ink. He's so young, and he loves tattoos.

I clutched my beer. I was half awkward and half at home. Lisa had rented out the whole bar, but she was nowhere to be seen.

● ● ●

Lisa worked hard to put herself through school. She worked hard to provide a decent middle-class life for her children. She studied at night, and worked, and studied some more. She put aside the wildness of her early life by herself, like an adult.

Lisa has a sweet twelve-year-old African-American son who likes chess and bird-watching. She has a dangerously beautiful fifteen-year-old blond daughter who likes risk too much. She is afraid for her daughter, and when she showed me the video of the girl attempting to back her car out of the garage when she was at work, I understood why. She did not have a license, and did not know how to drive.

It's hard, that single motherhood. She got the video because she put in a camera. She put in the camera because she was suspicious.

"I was like that, too, when I was her age," she said. "I know how her mind works."

"Really?" I said. "I'm shocked."

* * *

Lisa met the man of her dreams at a wedding. The man of her dreams is a little older. He's successful. He has an airplane and a business. He's divorced, with grown children. She showed me his picture. He looks like an ordinary man, clean-cut, fit, a little bald.

"He looks like a doctor," I said.

"What can I say," she replied. "It's true love."

On paper, your heart sinks. On paper, you wonder. To risk, to move away, to uproot, to change, for the tentative thing, the dream of happiness and togetherness and love. Children in school, children with friends.

Being an ER charge nurse is a good job. The money is decent. But it's hard. The responsibility is great, especially in a trauma center. You have to pay attention. You have to manage. You have to be approachable in your authority.

You have to use a light touch and a heavy one alike. Your phone rings all the time, and people come to you.

Lisa was making us wait.

● ● ●

Paul was there. He's a tech rather than a nurse. He's small, unassuming, often unshaven. He's the kind of guy people sometimes overlook.

But Paul is very smart. He talks to me, and he's a reader and a listener. His knowledge is broad and deep and more current than mine. He's a family man, and has a teenage son, and once he gave me an old skateboard, because he rides skateboards with his son in the irrigation ditches, where the concrete is smooth and beautiful and stretches for miles. I remember that, too.

He didn't dress up. He was like me, out of place in ordinary clothes.

We shouted at each other over the music for a while.

They were getting ready to dance. I could sense it, and I was nervous, because they were eyeing me like prey.

By then it was crowded.

● ● ●

Then there was a ripple at the door. Something was happening. People were walking toward the front of the bar, and I followed.

She had texted. She was about to arrive.

So we waited by the door, and just outside it, in the

warm night. The speakers were thumping behind us—
Kool & the Gang, the Bee Gees, KC and the Sunshine
Band, all of it. It was the music of my childhood, before
most of the crowd had been born.

There was an arrival song. I can't remember it.

Then a stretch black limousine pulled around the corner, eased down the street, and stopped in front of the bar.

People got their phones out. There should have been
flashbulbs. But flashbulbs are a thing of the past. Now it's
just the quiet little blink of the phones.

The door opened. The music was loud.

And then she got out, in her boa, and her feathers, and
her hat, and her sunglasses, and her minidress, and the
crowd started cheering and clapping and hooting, because
they were a little drunk, and they were so full of life, and
everyone liked her, and everyone was sorry to see her go,
and she looked like a movie star.

GLORY

HE WAS SO THIN. I COULD SEE THE BONES OF HIS CHEEKS AND his temples. He was sitting bolt upright, panting through pursed lips. He looked afraid, and he looked too young for emphysema, and his eyes were large and blue.

Beside him, on the chair in the lobby, was a cowboy hat.

I was walking by when I saw him.

"Do you have asthma?" I asked him. "What medical problems do you have?"

I dispensed with the niceties because I'd been going somewhere else.

"I don't have asthma," he said.

He was having a hard time speaking. The oxygen cannula hissed, cranked up from the bottle at his feet.

"How long have you been sick?" I asked.

"I don't know what I got," he said. "I been sick awhile."

"Have you seen a doctor?"

"That's why I'm here."

"Have you been losing weight?"

"Yes," he said. "I feel like I'm going to die."

I told the nurses to move him to a room.

"He looks worse now," the triage nurse said, apologetically. "I'm sorry I left him out there."

So I went to the doctors' station, and pulled up the X-ray that had been taken hours before. I looked hard at the film. But I didn't see much, just a little haziness in both of his lungs. His temperature was normal. He had a mild dry cough. He wasn't shaking with chills, or sweating. He wasn't wheezing, either. He could take a full breath. But he was struggling to breathe anyway, and he was so thin.

Right then it came to me.

They moved him to a room, and I told the resident to go see him. The resident was in there for a while. Through the glass I got a glimpse of him listening carefully with his stethoscope, pressing the bell against the man's chest.

The resident came back.

"What do you think?" I asked.

"He looks like he has pneumonia. He's hypoxic and needs to come in."

He was right. But he still missed it.

• • •

I could easily have been wrong. There were other possibilities. But I wasn't wrong.

The test I ordered came back an hour later, lit up red on the screen.

"How did you know?" the resident asked, staring at the computer. "That was a good call."

"It was just a guess," I said.

The resident is smart. He pays attention. Not much gets past him. At this point he doesn't need supervision at all, by anyone.

It wasn't such a good call, I wanted to tell him. *It was only the past. You're too young to remember.*

But I didn't say anything. I just let him think I was smarter than I was.

*   *   *

I went to medical school in the South. When you're a white man, there is often tension with black patients in the South. There is a certain formality, an unspoken edge.

But he was beyond any of that by then. There are moments, and times, like war, when identities cease to mean anything.

I can still picture him so clearly. He was about my age. How directly he looked ahead. I remember his thoughtfulness, and his recognition, and the sense of preparation within him, and how he spoke to me about it.

You see this sometimes. You see it in the oncology units. You see it in old people whose faculties are intact. I'm not sure that state of mind is acceptance. Nor am I sure that it is unafraid. But I do think it is part of the animal self, part of something ancient within us.

He was riding in a car through Greensboro, North Carolina, at night, in the back seat. He let his friend shoot him up, because he hadn't done it before and didn't know how. Everyone in the car used the same syringe.

Cocaine in the back seat, a rush of music and lights ahead. He was straight. He had a child, and a shit job at a chicken-packing plant. These things I remember. He hadn't graduated from high school. Maybe this story was true, or maybe it was something else. But it felt like the truth, the way he told it.

On paper, he might have been a simpleton. But he wasn't. He was intelligent and insightful and brave.

Sometimes you see that where you least expect it. Sometimes you find that in the back seat of a car, mainlining cocaine, as the music thumps, and the club beckons.

•   •   •

I remember the ward. I remember the hearses parked on the ground floor, and how every day the ward would give them business. I remember the fear we all had— fear of needlesticks, fear of the operating room, fear of casual sex.

That ward was an utterly terrifying place. Every single patient on it was dying. Each day brought new faces, and each day brought empty beds.

I remember a man, smiling in the morning as we rounded on him, dreamily laughing, confused, blind, with black spots all over his face, beaming at us.

He gave us chills, all of us, because he was so far gone he didn't understand anything, or see anything, and was babbling and smiling, light-skinned, African-American, with green eyes.

"I wonder what it will feel like," my patient said, talking about death. I've almost never had a patient speak like that, and so his words have remained with me, after all this time.

It was twenty-six years ago now. He died that spring.

● ● ●

"How old are you?" I asked the resident. I already knew, more or less.

"I'm twenty-nine," he said.

I thought about it. I thought about how he had been three years old that year, and how the generations pass, and how little I had to teach him. How quickly we go. We hardly cast a shadow. We pass through, we do what we can, and then we leave. The young do not look back.

So we went in there, the resident and I. We sat down and told the cowboy he had AIDS.

He took it well. He nodded. The nebulizers had eased his breathing. The antibiotics and the steroids were running in.

"It must have been my girlfriend," he said. "It must have been her."

We didn't press it. It didn't matter. No one cared what his story really was.

"Do you live here?" I asked him.

"I'm from El Paso," he said. "I was just driving through and it got real bad."

So we talked about the drugs that we were going to start, the drugs that he would need to take, and keep taking, that had come like the grace of God.

"You're not going to die," I said, because it was true.

THE TEACHER

The administrative offices of the Department of Emergency Medicine were once a morgue. The morgue has been moved to another building at the university.

The space is now a warren of windowless offices and conference rooms. It feels cool and corporate, like a hotel chain. There are vases with dried flowers on side tables, by couches no one sits on. There are prints and photographs on the walls—balloons over the desert, sunrise at Chaco Canyon.

If ever there was a place for ghosts, it is in these offices. Many thousands of people have been autopsied here. I think about that sometimes, in my tiny office, with the door closed and the computer on. But the truth is I've never felt anything, even very late at night when I'm alone in the empty building—no presence, no prickle on the back of my neck, nothing at all.

That morning, I went to check my mail. I stood beside

the photocopier in the mail room and went through the pile, flipping the pointless envelopes one by one into the trash.

There, suddenly, among the pharmaceutical announcements and credit card offers, was my high school alumni review. It comes twice a year, to my work address.

His obituary was on the front page.

• • •

When I think of him, I think of a tiny man with a gray beard, dressed in a dark and immaculate suit, pacing back and forth in front of the class, calling out numbers and letters.

"G," he would say. "X, 7, 1."

You could see his Japanese father in his face and his English mother in his hair. He wore it combed back, and his scalp gleamed through it. Sometimes he would call out words in his sharp voice.

"Cat. Dog. Evergreen. Quick. Brown."

At each word, there was an answering crash of electric typewriters. There were several dozen of them in the room, gray and battered, humming faintly. The students sat with their backs straight, hands poised.

Aside from his voice, and the typewriters, and the click of his heels on the floor, the room was silent.

I had a free period before lunch. I'd eat early, alone in the hall, sitting on one of the plastic chairs bolted to the wall, and watch the typing class through the open door.

I remember how I liked it, the call and response, and the perfect attention. It was almost forty years ago and my life was just beginning.

Then the bell would ring, and everyone would leap from their seats and become high school kids again. They'd start talking and laughing, and he would shout over their voices, calling for order as they milled around, and sometimes his voice would rise in a shrill and angry way as he walked through the rows, snatching the pages from their hands.

• • •

The school was a Canadian boarding school in Kobe, Japan. The students were from all over the world—Europeans and Americans, Indians, Japanese. The curriculum was taught in English, and everyone went to college—mostly American or European universities. The teachers were an odd mix as well—the handsome Spanish teacher, Mr. Hernandez, who was actually French; the math teacher, Mr. Patel, who was from India; the wretched British Roger Onions, who taught seventh grade and was covered with acne scars. There was a troubled American biology teacher, Mr. Rivers, who left his family for a beautiful seventeen-year-old Japanese girl who was his student, and an American ex-hippie, Mr. Williams, who said things like "far-out" and "grass," dated even then, and had a Japanese wife and a couple of kids. There was the headmaster, Mr. Albert, a strong, fit, confident, aggressive

American with a love of martial arts, who, though married with children, nonetheless imported his German mistress as his assistant and whom everyone was convinced, correctly I still think, was into S&M and bondage.

And there were my parents—my father, who taught junior high English; my mother, who taught fourth grade.

The school was set high on a hill overlooking the city.

The teachers were provided housing. We had a rented house by the train station, and it was fine. But all of the teachers aspired to one of the handful of homes that the school actually owned, on the hill near the campus.

They were made of stone, predated the war, and were built to Western dimensions. They had panoramic views of the sea. By Japanese standards they were mansions.

Victor Mihara lived in one of these houses. He was one of the most senior teachers at the school—he'd taught typing for thirty years by then. And so, eventually, as those above him retired, he reached the house on the hill.

He invited us over once, when we first arrived at the school. I remember the immense stiffness of the conversation, the heavy furniture, the heat of the late Japanese summer—there was no air-conditioning—as his heavy and nearly silent English wife served us tea. They had no children, and she did not work.

There was no sign of Japan in the house. There were paintings of the English countryside on the walls. There were overstuffed chairs and lace. There was heavy dark furniture.

My mother, in the heat and stillness of the parlor, as we all sat awkwardly, trying to make small talk, praised him for his elegant clothes.

"You always dress so well," she said, as a compliment. He smiled.

"Well, you see, my tailor is on Savile Row," he said. "And I have a block there for my shoes as well."

"How often do you go to England?" she asked.

"Oh, I return perhaps once a year," he replied. "I have family there, of course."

Yet who his family was remained unmentioned. He did not go on, but changed the subject to something else.

Later, on the way home in the taxi, my parents discussed him, as people do. They spoke of what an odd little figure he was, how he had lived in Japan for his whole life and spoke hardly a word of the language, could read no signs on the street. And they spoke of Savile Row, and pretentiousness, and working-class English accents, and how he was a typing teacher, and they weren't kind remarks.

• • •

I never took his typing class; by then I knew how to type. But he was my soccer coach, and that is where I knew him best. He would stand on the dirt field in front of the school with his whistle and his soccer cleats and his brown knobby legs and his beard, and from a distance he looked like a child.

Occasionally he would tap the ball to one of us with

his foot, but even that small act revealed his awkwardness. He would blow the whistle, and we would run a few laps around the field, and then we would scrimmage for an hour or so, and he would stand there watching, sometimes blowing his whistle, and that would be it.

He had been coaching the varsity boys' team for almost as long as he had been teaching typing, but he knew nothing about soccer. He would tap the ball, ungainly and slight, and jog a few steps one way or the other, and blow his whistle, and offer a few suggestions—*you must move wide, and cross the ball into the box; you should watch the eyes, not the feet, that is where the man will go*—that all of us ignored as we waited for him to blow the whistle so the scrimmage could resume.

Once, in the faculty-versus-student softball game, I watched him strike out, swinging the bat hopelessly at the slow pitches, thrown by a girl, smiling tightly with each miss, his eyes gleaming with fury and inadequacy, and I think it was then, as the crowd clapped politely, when I first realized how much he despised himself.

• • •

His death surprised me; he was the kind of man one expected, somehow, to live on into the far reaches of old age. And yet he did not; he is gone, slipped away as if he never existed.

Once, on the Shinkansen, north to Hiroshima for a soccer tournament, he gave us a motivational speech.

As the Japanese countryside streamed past, the train soft and effortless on the elevated track, through rain and clouds and brief patches of sun and the endless gray concrete of the towns, he urged us to try, to win, to fight, to work together as a team, and to remember what we stood for and show everyone what we were made of. The wretched Roger Onions, the assistant soccer coach, British and unpleasant, but nonetheless an excellent player who knew what he was doing as a coach, smirked slightly.

"Do it for me," Victor Mihara said, finally, nearly in tears. "Win for me."

His speech puzzled us, and we were quiet. Win for him? Our team was made up of American, Korean, Japanese, Dutch, and Indian kids. We didn't dislike him, exactly. We just didn't care about him, and that day, as Japan poured by through the immaculate windows, we stared at him in puzzled silence.

Later, of course, I understood. Victor Mihara, struggling for Kipling and Eton in his voice, wanted to be loved.

• • •

When you're young, you do not understand the humanity of adults. They might as well be another species entirely—adults, teachers, old people. You cannot see how weak they are, how small they can be, how full of desires and little hopes of their own and weaknesses that grow, rather than recede, as they age. The tailor on Savile Row, the block for handmade shoes, the pretense of Englishness, and yet there he was, cast out by two societies as a half-breed, from a time when that term flowered most cruelly, posing as an Englishman when he was not, sitting in his house on the hill, so careful to assert his ignorance of Japanese, so eager to claim he could not read the signs in the train stations, when in fact I knew he could speak Japanese perfectly. Once, in the locker room, when no one else was there, after soccer practice, when he'd thought everyone had gone, I heard him talking to one of the janitors.

In Japan, blood is everything. Traditional Japan is unapologetic in its contempt for any and all who are not Japanese.

Victor Mihara hated the Japanese. All of us could see it. And his suits, his house, his English wife, his pretense of ignorance—all of that was revenge, I think now, as if anyone cared.

Which, of course, they didn't.

• • •

There is hardly any sign of the war in modern Japan. The apocalyptic fires that left Tokyo in ashes, the hundreds of thousands of dead—all of that has been reduced to unspoken memory, a blank space. Even Hiroshima, or perhaps especially Hiroshima, has sealed over once again. Only the A-bomb dome remains, like a token, or a gesture.

The A-bomb dome is a concrete building with a steel frame. The bomb exploded directly overhead, and the shock wave struck the roof straight down. There was only vertical force, and the walls were strong, and so the building remained standing. Yet all of it is off, slightly bent and warped, not quite straight. If you look closely, you can see where the stone melted, ran down the walls in little rivulets, and dried again. The rest of the city looks like anywhere else—one could spend a month in Hiroshima, or a year in Hiroshima, a decade in Hiroshima, and never know anything had ever happened there. But I remember the human shadows anyway, burned into the stone steps of a bank by the flash. The steps had been excised, and displayed in the museum. The shadows were faint, and you had to look closely. I remember the fingernails full of blood vessels also, like black worms, from a man whose hand had been outside a concrete wall as he reached through a window. He had lived on, but his hand had been changed into something else, and for years grew blood vessels instead of fingernails from the tip of each finger.

* * *

The school where the soccer tournament was held was a brutal-looking place. It was all concrete and wires, like a concentration camp, with a dirt field. Most of the soccer fields in Japan at that time were dirt; grass was too expensive to maintain. The school was large, with thousands of students, and had one of the best high school soccer teams in Japan.

Our team was OK—most of our players were from soccer nations, in Europe and elsewhere, and we often played evenly against adult club teams, but Kansei Gaukin destroyed us every single year. The scores were humiliating—six to zero, seven to zero. Three to zero was a moral victory for us. Their players were fast and strong and disciplined and hard and skilled, and the military echo of imperial Japan was everywhere in that place. There was ambient ferocity as the horns blew. They wore black uniforms on the dirt field, and there was no color anywhere, and the locker room was cold and dismal.

● ● ●

Students in Japan are driven mercilessly. School lasts all day, and afterwards many students go to private preparatory classes to prepare for the all-important university entrance exams. Those exams are everything; a good score means one life, a bad score another. Every year, when the results are announced, there is a rash of suicides reported in the papers. One I remember especially—a boy hooked electrodes to his chest, connected to an alarm clock. He took sleeping pills and drifted off, and when the clock struck the appointed time, it killed him.

I remember thinking that it was a sophisticated way to kill oneself. It wasn't a bridge or a cliff; it wasn't a noose or a knife or a train. How badly could he have possibly done on that exam? The method radiated intelligence, calculation. His mother, the article said, was distraught. His father was not mentioned.

● ● ●

But that day, on that field, astonishingly, our team won. Soccer is lucky like that sometimes. Their keeper misjudged a long low shot on the gray cold day; it skipped off the dirt and went in. Then we held on, somehow. Their shots missed; we pulled all of our players back. It was a game that was a victory in name only, the first and only time we won against that team. The final score was one to zero. When the whistle blew at last, the crowd clapped politely under the gray overcast sky, in the cold, and our breath came in gusts. But on the train home, everyone was happy and proud.

No one was more elated than Victor Mihara. His face was lit with joy; his voice rose to a squeak as he congratulated us; he could hardly sit still; he was like a child, giddy and giggling—any pretense of dignity was gone. None of his teams had ever defeated Kansei Gaukin in more than twenty years of trying. Yet the moment had finally come, he had a victory at last, and looking back, it seemed obvious what he was struggling against—the cruelty in Japan and the cruelty in England, which even now remains just under the surface, where those who are weak and do not belong are cast aside and dismissed. And Victor Mihara was full of weakness. Weakness, and hatred of that weakness.

· · ·

I suppose the reason I think about him now is that I can imagine a little of what he must have felt. I, too, can feel the casual dismissiveness of the students, and the indifference in their eyes as he aged, as he became an old man and they were the same—young, with their lives just beginning, unaware of the fates that awaited them, without fear or recognition or awareness at all—just kids, in the passions and immediacies, as if there had been none before them, as if none would follow. Perhaps most of all I can feel the terror of Victor Mihara—in his example, in how he slipped away, in how little he mattered, in how few remember him, and how even fewer remember him with kindness or regret.

• • •

I've had many teachers, and I've seen obituaries before and simply turned the page. But on that day, I took the alumni review to my office, and sat down at the computer, and typed in his name. I'm not sure why, exactly. I suspect I was one of the few to have done so.

I did not learn his story. I didn't find out anything about his mother, or his father, or how he came to Japan, or why he didn't leave. I never learned, truly, who he was.

But I did find something nonetheless. It was a single reference in an obscure Canadian academic journal:

> Victor Mihara, a Japanese-Briton whose Cockney accent had been acquired in East End London and tested by war years in a Japanese prison camp.

It was not a story he ever told. No one knew that he had been in a prison camp as a boy, or what had happened to him there. The camps were terrifying, and a third of the prisoners died. I knew this, and as I looked at the screen on my desk, I understood for the first time that his resentments and his pretenses went far deeper, and were far darker, than any of us had realized. Now the details are almost completely gone—the story of the camp, of his mother and father and all the rest, the story of the undistinguished and the unremembered, present only in flashes, if we look closely enough.

But if I could, I would tell him how well I remember him now. I would tell him how clearly I can picture him, pleading with us on the train all those years ago, though we did not see it, and did not understand what he was asking of us, or why, because we were so young, and knew so little of the world, and the future was so far away.

THE LESSON

THEY ARE EIGHTEEN-YEAR-OLD KIDS FROM ALL OVER THE
state, from small towns, and the countryside, and the reser-
vations. They are applying to medical school directly from
high school, an eight-year program designed to keep future
doctors where they were born, caring for the people they
grew up with. It's difficult to attract doctors to these places.

So they file in, one by one, in their new suits and
dresses, blinking in the office, and usually their parents are
with them.

She came by herself, with her wide blue eyes and her
friendly face. Her dad was a rancher in the south, along the
Texas border.

She was nervous in her dress.

"So," I said, after a bit, because the interviews were only
a few minutes long, "tell me why you want to be a doctor."

She blushed and stammered, and said that she had always wanted to be a doctor, and that she really liked helping

people, and that she had worked in the summers as the receptionist in an urgent care here in the city, and that it had been an amazing experience.

I had read her essay about this. I had also read the letter of recommendation written by the owner of the urgent care.

"Don't worry," I said. "Don't be nervous. Let's talk about your essay."

But of course she was right to be nervous.

• • •

This is the story she told. She was working late in the urgent care, at the front desk, shortly before they closed. An elderly couple brought in a baby. The baby was only a few months old. He was their great-grandson, and they were watching him. They were in their seventies. They came to the desk with the baby.

Unlike emergency rooms, urgent cares are not required by law to see every patient who walks in the door. They ask for insurance status and payment up front. If you can't pay, you are turned aside. Urgent cares operate like any other business.

The elderly couple didn't have the money. So they stood there with the baby at the desk, and she didn't know what to do.

She went to the back and talked to the owner of the urgent care. A doctor with an eye for business can get rich that way, far richer than working in an ER.

In his letter, he called her a wonderful young woman, and said that she had worked hard in his office. He said she had been a babysitter for his children, and he could think of no better example of his trust in her.

She asked him to see the baby.

Nothing is for free, he said. Tell them to go to the ER.

She said they were old and the baby was really young. She was upset.

I'll tell you what, he said, after some thought. I'm going to teach you something. If you want me to see the baby, then I will. But I'll take the fee from your paycheck. So it's up to you. Nothing is for free.

She thought about it, and agreed. She would pay for the visit herself. He saw the baby. He diagnosed an upper respiratory infection. He sent the baby home. The elderly couple was happy. And then he docked her wages.

• • •

She told me the story again, in her own words, softly and proudly.

"What was the lesson you learned from this?" I asked her, after a moment.

She hesitated.

"I guess I learned that sometimes it's important to do the right thing," she said. "I think I did the right thing."

"How much did he take out of your wages?" I asked.

"A hundred and fifty dollars," she said. "But it was worth it. The baby was OK."

"So," I said, after a bit. "What about him? Did he do the right thing?"

"Oh," she said. "Well, I think he did. He was trying to teach me."

"Do you think he should have taken your wages?"

Again, she was puzzled.

"Well," she said, "it was my choice. He gave me a choice."

"But do you think he should have put you in that position?"

"I'm sorry," she said then. "I don't really know what you mean."

She stopped talking and looked down at her hands. I could see that she didn't want to question him or what he'd done, that she was loyal to him, and had not understood him in the slightest way.

A moment passed.

"Tell me about the soup kitchen," I said. "The one you volunteer in."

"I go with my dad," she replied, with relief. "On Sunday afternoon."

So that was the path I followed. Soon she was talking eagerly about the soup kitchen, and her father, and her life in the small town, and I let her go on for a while because I liked listening to her.

"I think you did the right thing with the baby," I said at the end of the interview. "I think it says a lot about you."

She looked so happy.

THE BOY IN THE RIVER

IT WAS THIRTY YEARS AGO NOW, AND FOR ME IT WAS THE beginning. I'd finished my first year of medical school, but that was all I knew.

He was an orthopedic surgeon, a Scotsman in his mid-fifties with prematurely white hair. He was distant and quiet, guarded in his way, and I didn't get to know him well despite the months we spent together. He said little about God, and he never went to church when I was there. He was the only certified orthopedic surgeon in a tribal homeland of six million people, and the kind of missionary who did his proselytizing in the operating room.

"What is that muscle, Mr. Huyler?" he would ask me.

I'd just taken anatomy, and so I was often right back then. I've forgotten most of it now, the origins and insertions of all the muscles, the many strange names of the bones.

Anatomy is ancient, and full of the past, and learning it is like learning a language that no one speaks anymore.

"Yes, that's correct," he would say. "And where does it attach?"

So I'd hold my retractor, stare down at the wound, and try to answer him.

His patients were always black and always poor. Few whites lived in the homeland, and those who did were all in the capital city, where they had a tennis club, a school, and a few leafy green suburban streets with walls around their comfortable houses. If there was tension in the air, I didn't feel it. The pace of life was leisurely, and even the crimes reported in the local paper—murders and burglaries and rapes—had an oddly casual quality about them.

The doctor and his family lived in a small house on a ragged hill, just a five-minute walk from the hospital and miles from the white enclave. I stayed in their spare room—a hut in the backyard. In the evenings, from the porch of the hut, the faint glow of white South Africa was visible on the horizon.

His specialty was tuberculosis of the spine. By then he had treated more cases than almost any other surgeon in the world. Each week emaciated, paralyzed figures arrived from all over the homeland. Their relatives brought them out of the bush to the city, carrying them in the backs of pickups, or holding them upright on crowded buses.

During those months, I watched him expose spine after spine and drain abscess after abscess—tuberculous pus, buttery yellow and white. Cockroaches sometimes streaked across the operating room floor at our feet.

Orthopedics is mostly about the lame. But for him, every case was life and death. Sometimes, they even rose and walked again.

I think there was something in that for him. They rose and walked again.

He worked six days a week, at least twelve hours a day, and he got up in the middle of the night whenever someone knocked on his door. That is what strikes me most about him now—the effort his faith required.

• • •

Every summer he took a short vacation with his family. He liked to fish, which he'd done as a boy in Scotland. A wealthy white South African friend owned a house and a deep-sea fishing boat on the coast a few hours' drive to the east. He invited me along.

He asked me to drive down by myself, in his extra car. The family van was full, and he would need to return early from the coast in the car. I could ride back later with the others.

So a few hours after they'd all packed up their van and departed, and I'd completed my assigned duties at the hospital, I took his Opel and followed his directions east toward the Indian Ocean. I drove nervously, on the left side of the road, without a valid license, through a strange and foreign land, and everything seemed exciting that day.

The road turned to dirt at the edge of town, and soon the land around me was barren reds and browns, dotted with herds of thin cattle and the round thatched huts of rural Africa. The road grew steadily worse. As was customary, people tried to flag me down, and finally, reluctantly, I stopped for a young woman with a baby. Her face was a mask of tribal scars, and she sat in the back with her child, head bowed, for miles. She spoke only once, asking to be let off. The child made no sound at all.

It was late afternoon when I reached the river. The road, now a dirt track, simply came to an end by a hut in a dusty yard, the wide brown river flowing by. But I saw the doctor's van in the corner of the yard, and the boy who came trotting out of the hut seemed to be expecting me. He held up both hands, the gesture plain: *Wait.* Then he jogged away from me down the path along the bank.

Nearly an hour passed. It was hot and quiet. I wandered around the empty yard and then walked down to the river. I could faintly hear an outboard motor in the distance. It grew louder, and then a speedboat came around the bend, turned hard, and sent a long silvery tail of water into the air. A blond boy stood behind the wheel, and his brother scrambled up to the bow and threw me a line. These were the doctor's youngest sons.

The sea was a couple of miles down the river. As we left the confines of the mangrove trees and moved into the wide salt flats of the estuary, I saw the whitewashed mansion high on a hill, and the deep, radiant blue of the

Indian Ocean beyond it. It was the only visible building, with a red roof and terraces and a thin line of wooden stairs leading down to the dock by the river's edge.

The dock was modern, all new concrete. When I jumped out of the boat, I realized that the boys did not intend to stop. They left me standing there at the foot of a hundred steps and roared off again. I watched them go, turning left and right, sending up sheets of spray. Before I'd climbed halfway to the house, servants had converged, smiling, for my bag.

They gave me a deck chair on the terrace in the sun, and a tall man of indeterminate age brought me a drink. The doctor, he said, would not be back for hours. *Are you comfortable, sir? Would you like something to eat?*

The terrace was warm and peaceful. I could see for miles—the ocean, the glittering estuary, the blue sky. It was a beautiful place, the kind of house that would have cost millions anywhere in the Western world. It was furnished plainly, with old couches and comfortable chairs, and decorated mostly with photographs of the owner's family—a lovely young woman, a couple of handsome children. The man I assumed to be the owner was much older than his wife. He was fat and bald with a kindly face. In the photographs they all looked happy.

There seemed to be at least a half dozen servants in the house. I could hear their voices downstairs, and the smell of cooking rose to the terrace. The servant who brought my drink pointed out the doctor's cabin cruiser a mile or so offshore in the open sea. A figure stood in the tower, just visible, looking, I realized, for the flutter of schools on the surface.

I lay back in the sun on the terrace, sipping the gin, and after a while I dozed off. I don't know how long I slept—maybe only a few minutes—before the outboard woke me up and the older boy raced up the stairs.

"There's a dead body," he said, breathing hard, barefoot, in pale green shorts. "They want a doctor. Are you a doctor?"

I said that I was not.

• • •

The body lay facedown in the mud a few hundred yards up the river, invisible from a distance—we had unknowingly passed right by it only a short while ago. Two of the servants were already there when the boys and I arrived. They'd pulled him out of the shallows onto the mud at the river's edge, and rolled him over on his back.

He was perhaps sixteen. He was very cold and very black, like a stump carried and left by floodwater. I knelt beside him. He had no pulse. The whites of his eyes were red, his pupils dark and empty. His blue nylon shorts were wet completely through and clung to the exact outline of his legs. The doctor's sons stood wide-eyed and quiet and kept their distance. I looked up at them.

"It's the cook's son," one of the boys said. "We know him. He has epilepsy."

"He shook," one of the servants said, pointing to the body.

The man did an imitation of a seizure, shaking his arms, contorting his face. A bamboo fishing pole lay half in the water, its restless float trailing out into the current.

"Yes," he said. "He did not come back. We searched for him."

I looked at the boys, the servants, all of them expecting me to do something. But the cook's son was dead, drowned in a foot of water. He'd been fishing at the edge, and had fallen in.

"We need to get him up on the bank," I said.

So we lifted his body out of the mud. It was an effort to carry him up to the grass. Our legs were soaked. It was hot.

We left him there, with one of the servants to watch him, because I didn't know what else to do. We went back to the house and changed. We waited for the doctor. I remember how I could still feel the body in my hands.

Nearly two hours passed before the doctor guided the gleaming new cabin cruiser up the river mouth to the dock. By then we were on the terrace—the boys, the servants, and the cook. Only the cook was crying.

She was the tall thin woman leaning against the wall, and her tears tumbled down the tribal scars on her cheeks, dropping off into the blue expanse of her dress.

As the boat approached, I followed the servants down the stairs.

"So you made it, eh?" the doctor called over the burbling motor as the boat eased up to the dock. He threw the bowline to the servants, then moved nimbly down the deck to the stern. He was sunburned and shirtless, he looked relaxed and content, and his tennis shoes were covered with fish scales—some clung to the hair on his legs, some to his socks. When he was on the dock, I told him.

"Good Lord," he said, shaking his head, squinting in the sun, silent for a moment.

A woman spoke behind me. It was the cook, the last down the stairs, asking the doctor to take her son to the mortuary.

"Please, sir," she said.

"Of course," he replied, looking at her, reaching out to touch her on the shoulder as she began to cry again. There was a silence, all of us staring at our feet.

"Do you want to come with us?" the doctor asked as she wiped her eyes with the back of her hand. But she shook her head.

"I cannot bear to," she said, like an Englishwoman.

A crowd had gathered by the time we reached the body again. A village was nearby, invisible from the river, and the news had traveled fast. The men and women stood in separate groups, talking quietly. By then it was twilight.

The two servants met the boat, held the line, and then stood in the shallows as the doctor and I lifted the body and carried it down to them. We had an audience, and we tried to do it well, but the body was loose and heavy, soaking our hands as we struggled through the high grass. Getting it into the boat was more difficult still. One of the servants stood in the boat, his arms wrapped around the body's chest, grunting with effort as we held the legs. Finally we slid the body, as cold as a wet mattress, up and over and into the boat.

The doctor sat in the bow, with the boy's exposed face between his feet, and we went off down the river toward the cars. The two servants cupped their cigarettes in their palms, the doctor's white hair rose up, our shirts flapped around our ribs, and it got cold quickly. The presence at my feet seemed immense.

No one was in the yard to watch us, so we simply dragged his legs through the shallows to get to the cars. The back seat of the van folded down neatly, and we laid him there. The men walked back and forth through the headlights, mooring the boat, and the doctor wiped the mud from the boy's face, his bare legs and feet, then threw the towel down toward the boat in the darkness.

There was an old green bedspread in the back of the van. We used it to cover him.

"I knew him his whole life," the doctor said, shortly, as I watched him. Then he shook his head, got in the van, and started the engine.

It was a long drive, well over an hour. The men sat cross-legged in the back and gave us directions. The body shivered and leapt on the rough dirt road, sliding into them, and they struggled to hold it still. The headlights were mostly full of dust. Occasionally, the men in the back said something to one another, but the doctor and I didn't speak. It was warm with the heater on.

The hospital with the mortuary was a small mud building, but it was open, and a radio played somewhere as I walked in alone. Only one nurse was on duty; I found her folding sheets, loading them into a cabinet.

"Excuse me," I said, and she turned around. "We've brought a body."

She studied me. "You need a death certificate," she said. "I can't admit the body without a death certificate."

"Where can I get a death certificate?"

"You have to wait until tomorrow."

I didn't know what to do. I went back to the van.

As I stepped out into the dirt yard, one of our companions stood in the headlights, urinating a long arc into the darkness beyond. The doctor regarded him impassively from the driver's seat, listening to me and watching the pale stream of urine in front of him.

"All right," he said. "Why don't you wait here."

I watched his erect figure disappear into the lit square of the hospital door. He came out a few minutes later.

"She just wanted some money," he said, with a quick dark smile.

The morgue was an outbuilding in the corner of the yard. It was filled with stainless steel rows of drawers, refrigerated, with dials; it was cool and odorless, astonishingly modern, like the clean innards of a dairy.

We carried him on the hospital's stretcher, wrapped in the green bedspread, through the lights to his drawer. As he lay there, the nurse entered the room. She held a Polaroid camera in her hand and bent over the boy, focusing on his face. Then the flash, reflecting off the metal cabinets, and then the drawer, closing very quietly and smoothly.

"For our records," she said, and we watched as the image rose out of the black film, his face taking shape, the corners of his mouth and half-open eyes, his white teeth, his hair dark against the sheet.

We climbed back into the van, and she followed, casually, waving the photograph in her hand to dry it. We watched her through the windows as she crossed the yard and entered the door, then closed it behind her.

Everyone was quiet on the drive back. The road was empty, and for a few minutes the doctor drove fast, taking the curves with a shudder. He liked to drive fast. But after a while he slowed down again.

"Sorry about all this," he said, as if somehow he were to blame.

We settled into an uneasy silence, with only the sound of tires on the washboards, and the occasional hiss of a match striking behind us. The van was full of smoke, but dust poured in the windows if we opened them.

Nearly home, we came to a gentle stop as a small herd of cows, looking very white in the headlights, drifted out into the road and turned their slow heads toward us. The doctor didn't blow the horn, but merely waited as the engine ticked over, and a few dim sounds from the fields came up, and then the cows continued across the road and disappeared.

• • •

I've never been back to Africa. Those boys are men, and
the doctor is gone now. His free clinic is closed and forgotten. The work he did has entered the past, along with
the photograph he received at the end, shortly before he
died. Prince Charles is pinning an award on his lapel, in
Buckingham Palace. It is a lesser honor, and no one will
call him sir. A flash, and a handshake, and on to the next
in line.

I've seen so many seizures over the years. There is a
sudden change. They look off into space. Sometimes they
cry out. Then an arm or a leg starts to twitch, and then
it's all over them, dark and otherworldly. After a minute
or two it stops, and they go limp. It takes many minutes for them to wake up. This was when the cook's son
drowned—not during the seizure but in the aftermath. I
did not understand that at the time.

An event like the boy's drowning is really a cascade of
small moments: what seems singular is manifold. The boy
goes fishing, and he has a seizure, but he falls backward
on the grass instead of forward into the water, or whatever drew him down to the water's edge—something he
saw, perhaps, out in the current—never appears. Or he's
not alone.

Tragedies are not entropy. They are competing forms
of order. I can get no further.

• • •

When I was younger, I looked at God foolishly, which is to say I dismissed him. I still do. But I am reassured by those who do not. True faith, in others, comforts me. It suggests that grace is human, that it is within us. To me, those who examine this world should know better. If there is a God, he is not here to save us. But some, somehow, choose not to know better. They can see the questions clearly, yet turn to the simplicities, and the mysteries, nonetheless.

I can still picture the doctor standing in the headlights outside the morgue. I can see him with perfect clarity. He was an intelligent man. He had given his life to this work, to this endless struggle, and now he is forgotten by all but the few. I think about him, a stranger, a man I hardly knew, because I am nearly his age now, and great amounts of time are gone, and yet it still feels somehow like the beginning.

I remember his presence beside me as we drove back to the house on the hill. I remember his sudden words also, after many minutes had passed, in the middle of that long night. I understood that he was speaking to himself, and did not reply.

"At least he was saved," he said, his eyes fixed on the dirt road in the dark. "At least there's that."

THE SLEEPER

FROM A DISTANCE, THE BUILDING COULD BE ANYTHING—
offices, or a factory, out in the scrub north of the city. You
can see it for miles as you drive down the two-lane road,
with mountains in the distance and the big sky overhead.
At night, lit up from one end to the other, it gleams like a
ship at sea.

But it's only up close that you can tell what it is—a hospital, built in the middle of nowhere, where the land was
cheap. From the windows of the higher rooms, you can
look a long way out into the desert.

The hospital is brand-new. It was built to lure the insurance of the suburbs. But the projections were off, and
the insurance has come in a trickle rather than a gush. So
other calculations have been made, and services have been
cut, and now the hospital feels shiny and clean and strangely
empty at the same time.

When I work there, as I sometimes do, I'm always wary.

It's part of the endlessly uneasy solitude of medicine in
small hospitals.

• • •

He'd driven in alone, a large man in his fifties, with a
belly and a white goatee and wide, faintly startled blue
eyes. He wore clean jeans and a T-shirt. Chest pain. But
he wasn't gasping or sweating or clutching himself. Instead, he looked nervous and shy, like a student suddenly
called on in class.

"I was on the job," he said, like an apology. "Something doesn't feel right."

A foreman on a road crew, a construction site on the
outskirts of town. He didn't go to doctors much.

They handed me his EKG.

• • •

EKGs are magical and uncertain. They're from another
era, when antibiotics did not exist and the heart seemed
endlessly mysterious.

In the deepest sense our lives are electrical currents.
Charged elements—sodium and potassium and countless
others—flow back and forth across membranes with impossible complexity.

The heart, unlike the brain, is simple enough to invite
understanding. It's dumb and brutal, charging itself up
again and again like a firefly.

All of our hearts beat the same way. The impulse begins high up, near our throats. It flows down through the

muscle, causing it to contract in a ripple rather than all at once. Blood is forced from the four chambers in an ancient and elegant order.

By the late nineteenth century the first tracings of the heart's electrical activity were made. That initial curiosity led to a great discovery: when heart tissue is damaged, electricity flows through it in different patterns. Those patterns are distinct, and leave tracings on a page that can reveal an astonishing amount of information to the practiced eye.

Ten electrodes are applied to the body in precise places. Each electrode measures the current between it and the other electrodes. Taken together, they record the currents of the heart in three dimensions. An EKG, most basically, is a two-dimensional graph of three-dimensional space.

After a while, when you've looked at thousands of EKGs, you forget the technicalities. You forget the answers to the questions they asked you on the tests. You just see the patterns, like familiar faces on the street.

I wasn't sure about his EKG. It didn't look quite right, yet it also didn't look quite wrong. It's another of the mysteries of EKGs: they are often opaque. They do not always reflect what they are supposed to reflect. Instead, they speak to probabilities rather than certainties, suspicions rather than knowledge.

*   *   *

I can remember cartloads of heavy paper charts brought hours late from the depths of medical records, like books in a library. All those rows of files, all those X-rays labeled with the Dewey decimal system and slid into thousands of paper sleeves—all gone, burned in incinerators, in the landfill, so many stories and terrors and struggles among them.

Now the records are only a few keystrokes away.

I entered his name and date of birth in the computer. I searched for him for a few seconds. But nothing came up.

Then a nurse cried out, and I turned.

He lay twitching and gasping on the gurney. A moment earlier he had been speaking to me.

Ventricular fibrillation is something you can tell at a glance. It's a lethal heart rhythm, a tiny sawtooth pattern on the screen. Instead of an elegant ripple, the current spins in wild little circles, and the heart stops pumping. You have to act immediately.

A normal heart beating in real time looks like a machine. It has a prehistoric quality, glistening with yellow fat, deep and dark and red. I've held a beating heart in my hand a few times, and I've always been struck by the immense impression of strength it gives—a knot of muscle bearing down again.

I've also seen ventricular fibrillation in an exposed heart. I've seen it in people who were shot and stabbed, when the surgeons have opened their chests in the trauma room.

The heart trembles like a fish left on the bank too long. Muscle, deprived of oxygen, shakes.

* * *

Defibrillation is an old trick, nearly as old as the EKG itself. It's become famous—"Clear!" everyone shouts, as the life-saving shock is given.

When the heart is shocked, all of its cells fire at once. For an instant, the slate is wiped clean. And then, one hopes, the inherent, mysterious, automatic nature of the heart will begin again on its own. It will charge itself up like an animal, and try to follow the well-worn path it always has.

If our hearts are too badly damaged, or too much time has passed, the shock doesn't work. Death comes quickly then, and it looks painless. Its speed is both terrifying and reassuring, because there is no time for fear, or plans, or regret. Just a strange fluttering sensation, and we're gone.

But if the damage to the heart is not yet too great, defibrillation can seem like the work of God. It comes down to the thinnest of lines, just a minute or two either way.

So we fumbled, and rushed for the defibrillator, and ripped up his shirt, and slapped the sticky electrical pads onto his pale hairless chest, and charged the capacitor as he gasped and twitched and his lips turned blue.

A shock from a defibrillator sounds like the crack of a tiny whip. It hardly seems electrical. The arms jerk; the body recoils. The body will do this for a while even after it's dead.

We shocked him, once. For a moment, the green tracing on the monitor above his head went flat, as all the cells in his heart fired together. And then, just like that, his heart started beating normally again.

The life flooded back into him. His face pinked up, and he began to blink. A few seconds later he was full of sudden strength, struggling to sit upright, ripping off the oxygen mask we'd put on his face.

It was only then that we rolled him out of the cubicle to the resuscitation room. Everyone—the nurses in triage, the nurses in the ER itself, and me—had been wrong. We hadn't sensed the urgency at all.

"What happened?" he asked. "Where am I?"

I didn't answer. I was paging the cardiologist. He'd looked OK, and the EKG had chosen, at that moment, to hide the truth rather than to reveal it.

• • •

At that time of day, during that particular week, there was a cardiologist in the hospital doing elective cardiac catheterizations. The crew was there, the equipment was ready, no one had yet gone home for the day. I didn't have to call for a helicopter and wait those long minutes for the rumble of rotors on the roof.

For the moment, poised between life and death as he was, that man had raw emotional power over us. He had everyone's complete and perfect attention, when you know the stakes are real, and they fill you up. Soon he would be forgotten, but that moment of power, when you are thinking of nothing else, feels pure and athletic and cleansing. It has a terror to it, but also a kind of joy, when the world is heightened and full of significance.

We gave him a drug that quiets electricity in tissue—amiodarone. It works a little bit. We gave him drugs that resist blood's endless compulsion to clot—aspirin, heparin. They work a little bit, too. But he was having a heart attack in front of us nonetheless: an artery was plugged, and his heart could descend again at any moment. What he needed was an open vessel, and fresh blood to the muscle.

Another EKG. I stared at it. And it still didn't reveal its secret.

●　●　●

The cardiologist walked in a few minutes later, introduced himself, and listened for perhaps thirty seconds. He glanced at the EKG, and he glanced at the rhythm strip, and then he shrugged.

"OK," he said, casually, to no one in particular. "Let's take him."

Take him—it's one of the more resonant phrases in medicine. It means action, doing rather than thinking or debating.

The man lay there on the gurney, listening to the cardiologist go through the details in his practiced way—a catheter into his heart, to open the blockage we suspected was present. The risks, the benefits. The form to sign. The minutes were passing.

He hesitated.

"I don't know," he said. "I want to talk to my wife. Do I really need that?"

He scratched his chest, nervously and absently. I remember that—a man with blue eyes and short gray hair and a white goatee, scratching his chest. Why the gesture stood out to me I don't know, but it did. It seemed human, I suppose, something I could understand. The electrodes itched, and he scratched at them, without thinking.

I realized that he did not know how much danger he was in. Everyone else in the room understood this perfectly. Only that man, bewildered, could not quite believe what was happening to him.

"There's really no time for that," the cardiologist said, and we all looked at the man in silence.

When that many eyes are on you, and everyone is waiting, and there are needles in your arms and lights above you, and drugs are running into your veins, almost no one can refuse. The fullest weight of authority is upon you.

"OK," he said, finally, and then he signed the form, and looked truly afraid for the first time.

"I want to call my wife," he said. But no one answered. They just rolled him away, down the hall to the elevators, at a speed that suggested urgency without panic. That speed is distinctive and unconscious in hospitals. It's a brisk walk, and you can see it immediately.

My shift was just beginning.

● ● ●

Progress in medicine follows a predictable path. The first step is diagnosis. The second is the performance of treatment. The third is treatment that works.

The first two, inevitably, occur together. But we wait for the third, sometimes for hundreds of years.

So much of what we do is ritual. People live, and die, as they always have. The minor illnesses of life get better on their own. The terrible ones get worse. There are only a few moments, and a few conditions, where medicine earns the faith we want to place in it.

But the knowledge itself is mystical. Blood, for example, changes color with every heartbeat. Venous blood, its oxygen consumed by the body, is a deep bluish-red to the naked eye.

The right side of the heart pumps venous blood back into the lungs, where it is transformed by the air we breathe. When blood pours out of the lungs into the left side of the heart, then flows through the aorta and its tributaries, it is scarlet again, full of oxygen and power.

The first of the aorta's tributaries are the coronary arteries: two little holes in the aorta, just above the aortic valve, one to the left, one to the right, each supplying a net of vessels that wrap the muscle of heart, and keep it bathed in endlessly scarlet, oxygenated blood.

Those vessels are hardly larger than a straw. They are always working, because the heart is relentless and thirsty. And so they are vulnerable, and when they become caked by time, by modern life, as they narrow with plaque, which feels hard to the touch, like gravel rather than flesh, our lives begin dangling from them. If the plaque cracks, and is opened by the current, the blood around it responds as it does to a wound.

It clots. It plugs the vessel. And in an hour or two, all the downstream muscle will be dead.

● ● ●

When I was a child, heart attacks could only be watched. Drugs that did not work were given. Blockages could not be opened. Hearts died, sometimes quickly, sometimes slowly, as everyone stood there, adjusting irrelevant drips.

But then the third step came.

Cardiac catheterization is astonishing and delicate and beautiful. The thinnest of catheters, little prehensile tails with intelligent tips that bend and twist, are threaded through needles into the arteries of the wrist or groin. They slide effortlessly through them, into the aorta, and finally up to the heart itself.

The heart beats, shadowy, on a black-and-white screen. The catheter is visible, inching forward, waving like a tendril in the current. The cardiologist is quiet and intent. All of the movements are gentle.

The catheter rises inside the aorta until it seems to be above the heart. But then it begins to bend of its own accord, following the arch of the vessel, descending toward the aortic valve. You can't see the valve on the screen, or any of the details, because X-ray images are only moving shadows, confluences of light and dark.

The cardiologist injects dye from the tip of the catheter. It looks like squid's ink on the screen. And then, a beat or two later, the coronary arteries leap out of the gray background as if lightning has struck. The vessels, for an instant, are exquisitely clear.

They freeze the image. They study it on the screen. The vessels look like root systems, deep and secretive.

That idea—the notion that a piano wire, threaded in from the groin, could stop a heart attack as it occurs—is amazing. The idea itself is simple. But it seems so alien and unlikely nonetheless.

●   ●   ●

Later that night, I looked him up on the computer again. I'd heard from the nurses already, but I wanted to see the note, in black and white. The cardiologist had gone home many hours before.

And there it was—a branch vessel on the back of his heart, precisely where the EKG was least accurate. A vessel little more than a thread, but it was enough. The dime-size piece of fibrillating muscle acted like a spark in dry tinder. All of the normal muscle around it lit up in an instant and began fibrillating as well, a descent into entropy that ended only in luck. Had he not come in when he did, had too many backs been turned, he would have been dead.

The cardiologist slid in the catheter and placed a tiny metal straw known as a stent across the blockage. The blood flowed once more.

The procedure took twenty-three minutes from start

to finish. But the dime-size circle of dying muscle became pea-size, and then hardly more than a pinprick. His
heart had barely been damaged.

* * *

I hesitated at the end of my shift. It was late, after two in
the morning, and dark, and I almost went out to my car
and drove home, because that's part of it also—leaving it
behind, the good and the bad alike in their facelessness.
But that night I indulged myself, as if I were young again,
and walked into the main hospital, the halls lit and empty
at the hour.

The ICU was on the fourth floor. There were only
a couple of patients, and a couple of nurses on duty that
night, when so little was happening. I waved my badge at
the detector, the door opened, and I walked in.

A hospital ward in the middle of the night has a kind
of softness to it. The lights are down, the sick are sleeping
around you, the bustle of the morning is a few hours off.

The lights in his room were low, the flat modern mon-
itor above the bed lit up, red and blue, digital and silent.
The curtains were open because he was being watched, and
I could see him through the glass door in the twilight.

His head was on the pillow, his chest rising and falling
easily. All of the numbers above his head were normal.

I watched that man, and understood that he would
never know I'd been there, and I suppose I was there more
for myself than for him, but as I looked at him, I felt no

sense of pride or ownership or power. I didn't know him, and would never know him. But as I looked at him as he lay sleeping, he seemed miraculous to me. He was alive because of distance and science and curiosity, because of the rational mind and the cold eye that we've learned to cast. It wasn't warmth, or empathy, or faith, or tenderness that had saved him. It was luck, and the idea of the body as a machine. That, and the folly of this hospital, built in the scrub a five-minute drive from the roadside where, on the given day, he'd worked.

Glory, like failure, like so many of the black stories, is private and small in medicine. But there are moments of breathtaking greatness also, and they, too, pass unspoken like ordinary days, and that night I was very aware that I was looking at the product of greatness, and that I shouldn't deny it. I thought of him waking up in the morning, blinking at the view, and I know from experience that it might seem like nothing to him, that he might easily fail to understand how close he'd come, that like a child he might shrug it off, and never grasp, or wonder, or recognize, and that when his ordinary life continued, it would seem as ordinary as ever. But these are the moments we cling to, and, as time passes, must remind ourselves to cling to. In a few hours that man would walk out of the hospital under the big Western sky with his life before him again.

As I drove home, I turned up the radio to keep myself awake, and I rolled the window down. It was a warm night, the stars were out, the hospital was lit up behind me in the mirror until it receded, and then I pulled onto the secondary road, and passed the Denny's and the Taco Bell, until finally I was on the interstate itself, another set of headlights in the stream.

THE MIRROR

WE REACH HER AFTER ROUNDS. SHE'S BEEN WAITING ALL night. I've just come on.

The intern tells me about her. She was slashed in the face with a broken bottle, by a man, after a party. It's a bad laceration. But otherwise he thinks she's fine.

So I walk in with him to where she waits.

She's young, Native American, a little heavy. She looks exhausted.

On her left cheek is a gaping, jagged wound at least five inches long, extending from her eye almost to her mouth. It's not bleeding anymore. I can see yellow beads of fat in it. Her cheek is swollen and bruised. The skin dangles. Her eye regards me above the wound. It's brown and clear and looks undamaged.

I introduce myself and sit down, and we talk a bit. The dangling skin shakes as she speaks.

I know the story already. Alcohol, an argument, the particular darkness.

There's blood on her shirt and blood in her hair. It's dry. I ask if she was raped, though not in those words.

"No," she says. "He just cut me with the bottle."

"Is he your boyfriend?"

She nods, acknowledging.

"Has he done it before?"

She hesitates.

"Never this bad."

She's not crying. She looks calm, and sober, and still. It gets busy in the morning.

* * *

I ask the intern how many wounds he has sutured. I know the answer before I ask the question.

"I've done a few," he says. "Maybe five or six."

Wounds like this take a long time. Your phone rings; the pagers go off. There are others to see, others who are more urgent. This is why she has waited, because wounds like hers can wait.

I can't let the intern suture this wound. She's a young woman, in her early twenties. The scar is waiting for her. She will have it for the rest of her life. It will be there in the mirror when she is eighty years old. The scar will be there long after I am gone and forgotten.

I think about that as I get ready, as I ask the tech to numb her and clean her, and get me a laceration tray, and gloves, and sutures.

When you're young, you do not think of this work as outlasting you. You do not think of a laceration as a legacy.

• • •

The tech has done his job well. She is lying flat on the gurney, with her head turned slightly to one side. She blinks. The wound is clean and shocking. It's bleeding now, a little. I look carefully for pieces of glass, but see none.

If I stand and bend over her, I know my back will ache. So I sit down beside her. I open the laceration tray on the table next to me. I put on my gloves. I take the sterile drape from the tray. It is white cloth, with a hole in the center.

I cover her face with the drape. The wound is the hole in the center. I adjust the drape so her mouth is open to the air. I want her to breathe easily, and lie still.

"Are you OK?" I ask, and her head nods, under the drape. Her eyes are covered like a blindfold.

"All right. Just lie still. Don't reach up and touch anything. Tell me to stop if you need to, and I will. Can you feel that?"

I touch the edge of the wound with a needle. It releases the tiniest speck of blood. But she doesn't move.

"No," she says.

So I begin.

• • •

The suture I use is very fine, because it is her face. I'm not a plastic surgeon, but I've sutured a lot of wounds by now. I've sutured hundreds and hundreds of wounds.

Mostly, you are quick. Mostly, you don't have to be

so careful. The body heals wounds; it is not the healing that is in question. It is the scar.

My eyes are not as good as they once were. I am near-sighted, and suddenly I am a little farsighted as well. I do not need bifocals yet. But bifocals, I suspect, await me.

So I have to keep the wound at a precise distance to see the thread. It is as thin as an eyelash. I can feel my age in that distance, because once there was no distance at all. The overhead light shines, and the silver needle driver glistens. My hands shake, but very slightly, just enough to see. The needle is the faintest crescent moon.

Suturing a wound can be a meditation. Suturing a wound can be beautiful. There is a rhythm to it, especially with a big wound. You ease the point of the needle in. The needle is curved. You roll your wrist, then look for the rising dimple in the skin on the other side. You snap your wrist a little then, and the bright point leaps out. You release the needle driver and grab that point. You roll your wrist again to follow the curve as you pull the needle free, trailing its blue filament. You tie the knot. You cut the thread. You pick up the needle again. Blood rises in little points.

When you are careful, you pull the threads together before you tie the knot. The edges of the skin converge. You study the position, and decide. Sometimes you have gone too deep, or not deep enough. Sometimes you are too close to the edge of the wound on one side and too far on the other. So you don't tie the knot. You pull the thread free and start again. It takes longer.

You are looking for order, for an exact and stately progression. You want the wound to come together gently. You want the edges to rise against each other, just a little. You cannot tie the knots too tight. But they must be tight enough.

When you have sutured hundreds of wounds, you sense these things. You don't really think about them.

But for her, on that day, I do think about it. I think about the bottle, and how he'd slashed her. I realize that I'm trying to give her my absolute best.

• • •

She has an itch. I stop, and put my hands in the air.

"OK," I say, and she reaches up, tentatively, under the drape, and scratches the tip of her nose.

"We're about halfway done," I say, and she nods, because I've told her not to talk unless she needs to. I am working close to her eye. I have to be careful. I have explained this. So she is careful. She, too, is trying her best.

Her hair is shiny and black and spills out from under the drape onto the sheet.

I continue. I find the rhythm again. The phone on my hip rings, and I ignore it.

I know they are waiting for me, the residents and students. I know the charts are piling up, that we are falling further behind. But I feel no urgency. I don't care at all.

Point by point, blue knot by blue knot, the wound

comes together. The tissue around it is bruised, so it is more difficult.

I'm sitting so close to her that our breath mixes in the air between us. But I'm hardly aware of her. I'm thinking only of the line of sutures, and the spaces between them.

• • •

Finally I'm done. I put the needle driver down in the tray. My back is stiff. I don't know how many minutes have passed. I lift the drape from her face, and she blinks in the light.

"OK," I say. "I'm going to wash it off a little bit."

I use pads of white gauze soaked in saline and hydrogen peroxide. I wash her cheek as she lies with her eyes closed. Blood has risen in pinpricks, here and there. The hydrogen peroxide foams when it meets the blood.

It feels tender, like something one might do to a child. I dry her cheek with the gauze. I am old enough to be her father. I am easily old enough. Then I stand, and turn away, and put all the instruments into the sharps container.

She sits up on the gurney behind me, and I turn toward her again. For the first time, I see her from a distance.

In a few minutes, the social worker will come. In a few minutes, we will ask if she wants to talk to the police, ask if she has a ride, ask if she has someone to stay with or somewhere safe to go. We will give her referrals, and suggestions, and instructions. Then we'll send her home.

Her cheek is swollen and dark where he struck her.

But the wound is gone. In its place is a line of tiny blue sutures stretching from the outside corner of her eye all the way down to her upper lip. They look as good as I could ever make them. They look like something that can never be taken from me.

She reaches up and touches her cheek.

"Do you have a mirror?" she asks.

TIME

THEY ARE NOT SURPRISED TO SEE ME NOW. THEIR EYES DON'T widen when I walk into the room. I can feel my life passing.

"I'm Dr. Huyler," I say. "I'm the supervising doctor. I'd like to talk to you a bit also if that's OK."

So I ask the same questions again, after the residents have seen them.

Sometimes they're annoyed.

"How many times do I have to tell you people the same story? I just talked to the other doctor. Do I really have to tell you again?"

"Yes," I say. "You have to tell me again."

I ask them because their stories change. I ask them because sometimes what I hear and see are not the same.

There is safety in this repetition. Each story is a point on the graph. Each test is a point on the graph. We are drawing a line. The line will point the way.

But I can't explain this to them. They will never understand.

• • •

A gunshot wound, a little flash through the muscle of the thigh, missing everything else. Two blue holes, releasing little watery threads of blood that run down both sides of his leg. He's young and walked in, 9 mm, a party.

"You're lucky," I say, and he nods because he's fine; the wound will heal in a few days and be sore for a few weeks, but that's all. The bullet slipped through and did not expand—the cheap ammunition of the gangs, the cheap ammunition of the kids in cars.

We wrap his thigh with a roll of white gauze. He thanks us and walks out, carrying the crutches we've given him in one hand. He's limping only a little.

One story, I think, or another.

• • •

The trauma pagers go off, but it's nothing. A drunk with a laceration on his head. We wave him through, and wave them off, because he's awake and talking, and doesn't need the trauma team.

When I was young, my dreams were great and incoherent. My faith in them was vague. But I felt them pull me forward with their power. Power for what, I'm not certain. Power for whom, I'm not certain.

She's old and frail and speaks in a quivering voice that makes no sense.

"I think she has a UTI," the resident says. "But we're going to have to cath her."

"OK," I say. "Then cath her."

So they do. She howls, and waves her thin gray arms. She's too confused to urinate for us on command, and she can't walk anyway. She wears a diaper, and she's ninety-three. The nursing home has sent her for a fever.

●　●　●

I wonder why I've spent my life this way. It was an unlikely path for me. But each life must be spent. We can't stop, or wait. We have to spend it all.

It's too late for those questions. Put them aside, I think. Look forward, with your eyes up. There is freedom in this, and clarity. You can see the end. But you can also see the beginning. On the right hand is the father, and on the left hand is the son.

"He's ready," the tech, Randi, says to me. She has brown eyes and she bats her lashes at me deliberately. It's a game we play.

I smile, and go see him. He's the drunk with the laceration on his head, the one we waved through the trauma room earlier. He's snoring.

She's trimmed the hair around the cut, and washed it out, and it's numb. I put on a pair of gloves and probe the wound. The Q-tip goes down deep, and suddenly I can see bone, yellow-white at the bottom of the well. I feel for a fracture with my fingertip, but the surface of his skull is smooth, and the head CT was normal anyway.

I staple the wound because a scar doesn't matter in the scalp. It takes about ten seconds, and each of the seven staples I put in feels clean and exact. I toss the disposable stapler in the trash, and peel off my gloves.

The irregular track of staples shines in the bed of his black hair, mixed with dark clots of blood. It's good enough. He sleeps on.

• • •

When the eye is too cold, when it's down near zero, it looks through everything, and everything seems the same. This life, or the next. This man, or another. This woman, or another. This child, or another. But everything is not the same, and you have to see that, too.

Medicine isn't answering their pages again. So I walk over to their room, where they sit beaten and defeated by the night, hunched over their computers. There are only two of them. They are trying to keep up, and sometimes they are overcome by helplessness. The pagers ring on their hips, and they hit the buttons and ignore them. They know, eventually, that we will come find them.

"Answer your pages," I say, because I'm sick of it. It's their job to answer, just as it is mine to call. *You think you are beaten and defeated?* I might say. *Well, you're not. You're just tired. This will all be over soon.*

They look at me. I can see their resentment. The young are unused to the naked expression of authority. They don't like it. They don't feel that I can tell them what to do. It's not the same as the past, when I was their age. I don't care. I know what it is like to be in their shoes. But they still have to answer, and everyone knows this.

• • •

Soon, I know, I should stop. How long, I can't say. I have a while; I have a few more years. But I'm not so far from leaving now.

He's vomiting blood, but not that much. He's shaky, and trembling, and I can see the story before me again. Empty bottles, a motel room for the night, because he got a little money somehow.

Sometimes I think I've given great bloody pieces of my life to people like him, and it feels like a tragedy for us both. At other times, I think I'm doing imperfect work that must be done in a decent world. I think of the floors that must be swept, the dishes that must be washed, the meals that must be cooked, and the houses that must be built, and how all the rest is commentary.

So I don't move him to the resuscitation room. I keep him where he is, by the doctors' station, and we give him the drugs, and watch him out of the corner of our eye, and after a while he stops vomiting, and his heart rate comes down.

"I don't think he needs the ICU," the resident says.

I agree.

• • •

Then the sun is rising, and it's time for rounds. The new team is here. Some of them have damp hair from their showers, and some of them smell like coffee, and they all smell like fresh air.

They are sharper than we are. Their questions make more sense. We are tired. My teeth feel gummy and my back aches. I have a few more millimeters of stubble on my cheeks, and if I let it grow, it will be gray.

We print out the list and hand it around. Then we start, one by one, plan by plan, story by story. There are at least thirty names on the list.

It's so easy to forget. It's so easy to let a fact slip by. So I gather myself and focus on the names, as if each one is a question on a test. In years past, I could do this. I can do it still. I can keep them in my head. I can remember the labs, and the findings, and the rest. But I use my notes more. It used to be a small point of pride for me: *Do not use notes, because then you forget. Keep them in your conscious mind.*

Of course, you keep them in your unconscious mind as well. They remain there, doing their work. This was something I did not fully understand.

The new attending is asking if antibiotics have been given.

"I think so," I say, and look at the resident. He checks the computer.

"Sorry," he says. "My mistake." He orders them.

But it's my mistake also. I should have caught it, because that's my job. It's nothing fancy. It's checking the boxes, and checking them again. The resident has been working hard.

"Sorry," I say to the attending. Maybe he'll use it against me, maybe he won't. Time passes.

"It's all right," he says. "It looks like a busy night."

The truth—it was an average night. It wasn't that busy. But no one has died, and no one, I don't think, has gone home to die.

• • •

Then I'm out in the open air, and the sunlight strikes me, and I can feel my body respond. I put on my sunglasses. The sun is low in the sky.

I cross the street. I feel as if I can see every grain on the asphalt before me. I walk across the grass, by the bus stop, and into the parking garage, where my car waits in the shadows.

I get in, and start the engine, and wait a little while, blowing on my hands, and the radio comes on. I pull out of the garage and out onto the street, heading east.

The winter sun in New Mexico is breathtaking. Driving as it rises is dangerous here. If you let it, it will fill your windshield with white hot light, and blind you in the mirror with its power. I've had to pull over and wait many times until I can see clearly. But I lift my hand, and shield my eyes, and keep going, because I want to get home. I want to make it to the corner where I turn.

My house is a short distance from the hospital. I've lived in this mysterious city for twenty-five years. I know that when I make this drive for the last time, no one will give me a backward glance. I'll just be gone, like so many others before me, and so many others to come. There is a kind of comfort in this recognition. There is a liberation to anonymity, as life ripples through you, and continues on. I did not understand this as a young man.

I turn left at the light, then go down the street past the new bus stops, past the restaurants and stores that seem to change every few months, closing as the rents rise, opening as they fall again.

I turn right, then left, then right again, onto the street. I pass the modest houses, one by one, and pull into the driveway of my own. It, too, is a modest house. I've chosen it also, because the extravagances don't suit me.

The garage door is open, and a plume of exhaust rises from the new car as it warms up. They're running late. I thought they'd be gone, because the day is starting.

I pull into the driveway and stop, and I feel stiff, and tired, but clean sheets and a quiet dark room await me.

I turn off the car just as the front door of the house opens. There he is, my beloved son. He's in the doorway with his backpack, looking back into the house, and he hasn't seen me. He has light brown hair, and blue eyes, and square shoulders, and he's no longer a child. He is taller than his mother, and soon he will stand evenly beside me. He is waiting for her. She is taking him to school. His memory is better than mine now. He can look at rows of numbers and remember them. I could also, once.

I can see his breath in the cold air of the morning.

ACKNOWLEDGMENTS

TO THE MANY WHO HAVE HELPED ME SO MUCH WITH THIS book—Chris Bannon, Douglas Binder, Will Blythe, Helena Brandes, Laura Brodie, Jennifer Brokaw, Jim Fleming, Jim Garland, Elizabeth Hadas, Deke Huyler, Marina Huyler, Kimberly Meyer, Michael Mungiello, Holbrook Robinson, Kenneth Rosen, and David Sklar—thank you.

I'm especially grateful to my editor, Jennifer Barth, and to my agent, Michael Carlisle, for their continued support over these many years.

Particular thanks, also, to the MacDowell Colony, where the bulk of this book was written, and to the Corporation of Yaddo, where it was first begun.

Finally, I would like to thank the Department of Emergency Medicine at the University of New Mexico School of Medicine, where I have spent my working life.

Individual stories, in altered form, have appeared in the following publications:

The American Scholar ("The Sleeper," "The Boy in the River")

Byliner ("The Wedding Party")

Columbia: A Journal of Literature and Art ("The Good Son")

New York Daily News ("War," "The Gun Show")

TONIC/VICE ("A Visitor," "The Horse")

Though this is a work of nonfiction, details have been changed to protect patient confidentiality.

ABOUT THE AUTHOR

FRANK HUYLER is an emergency physician in Albuquerque, New Mexico, and the author of *The Blood of Strangers*, *The Laws of Invisible Things*, and *Right of Thirst*. His poetry has appeared in *The Atlantic*, *The Georgia Review*, and *Poetry*, among others.